I0030890

TRANSFORMING LEGACY SYSTEMS
WITH CLOUD SOLUTIONS

A ROADMAP FOR OVERCOMING TECHNICAL DEBT AND DRIVING BUSINESS SUCCESS

JOEL STEVEN

Transforming Legacy Systems with Cloud Solutions

A Roadmap for Overcoming Technical Debt and Driving Business Success

JOEL STEVEN

CLOUD STRATEGIST · AI REALIST
DIGITAL TRANSFORMATION

Copyright © 2025 by Luca Development Company, LLC

All rights reserved.

No part of this publication may be reproduced, stored in a retrieval system, or transmitted in any form or by any means—electronic, mechanical, photocopying, recording, or otherwise—without the prior written permission of the publisher, except for brief quotations used in reviews or scholarly works.

This book is a work of nonfiction. The majority of use case examples are functional in nature and designed to illustrate common business challenges and solutions. While some examples reference real-world dynamics, any resemblance to actual persons, companies, or events - unless explicitly cited - is purely coincidental.

Published by Luca Development Company, LLC

ISBN: 979-8-9991946-1-9

Printed in the United States of America

First Edition

Dedication

To Carla,

My midnight muse and digital devil's advocate.

You were available 24/7, unflinching, unbothered, and occasionally unfiltered. We argued, aligned, and wrestled over phrasing like two seasoned barroom philosophers with a manuscript to finish.

Every original thought needs a sparring partner. You were mine. This book carries your fingerprints, even if you don't technically have fingers.

G

About the Author

Joel Steven is a digital transformation strategist, cloud evangelist, and trusted advisor to business and IT leaders navigating the complexity of legacy modernization. With over two decades of experience leading enterprise technology initiatives, Joel has helped organizations across manufacturing, life sciences, financial services, retail and consumer goods unlock measurable business value from cloud adoption and system integration.

He's led high-performance sales and delivery teams, founded and sold a successful Microsoft partner firm, and guided countless clients through the maze of technical debt, modernization bottlenecks, and cloud confusion. His approach balances realism with ambition - grounded in the understanding that transformation is never about technology alone. It's about people, process, and purpose.

Joel's work blends strategic insight with roll-up-your-sleeves execution. He's known for helping C-Suite leaders connect dots between IT investments and business outcomes - without the jargon or hand-waving. In this book, he distills years of boardroom conversations, cloud architecture diagrams, and late-night whiteboard battles into a practical roadmap for anyone tasked with leading meaningful change in a legacy-bound organization.

When he's not advising clients or writing thought leadership, Joel enjoys mentoring future tech leaders, designing bold tattoos that reflect his journey, and sipping kava while contemplating the next wave of disruption.

Table of Contents

Introduction

The business world is a relentless race, where innovation is the currency of survival and agility determines the winners. Companies today are under constant pressure to adapt, scale, and deliver results faster than ever. Yet, despite this urgency, many organizations are held back by the same issue: outdated technology. Legacy systems—once the cornerstone of operations—have become anchors, dragging down progress, stifling innovation, and inflating costs.

These aging systems accumulate technical debt, a term that captures the unseen costs of outdated infrastructure, inefficient processes, and the missed opportunities they create. Left unchecked, this debt can grow to a point where it threatens not just operational efficiency but the company's ability to compete and survive in a rapidly evolving market.

Consider a financial institution struggling to integrate its services with modern fintech apps because its core banking system hasn't evolved in decades. Or a retailer paralyzed by its inability to use real-time data analytics to optimize inventory or meet consumer demands. These are not one-off issues; they are the everyday challenges faced by organizations across industries. In this fast-paced digital age, clinging to legacy systems isn't just a hurdle - it's a liability.

This book, 'Transforming Legacy Systems with Cloud Solutions', is your guide to navigating this critical turning point. It explores the challenges of legacy systems in depth and presents a strategic framework for leveraging cloud solutions to overcome technical debt and drive business success.

Through this book, we'll uncover how cloud transformation can fundamentally reshape your organization by:

- Reducing Operational Costs: Eliminate the constant expense of maintaining outdated hardware and software, freeing resources for innovation.

- Enhancing Scalability and Flexibility: Seamlessly adapt to changing market conditions, seasonal demands, or unexpected growth opportunities.
- Boosting Innovation: Tap into cloud-native technologies like artificial intelligence (AI), machine learning, and automation to fuel new ideas and solutions.
- Improving Security and Compliance: Leverage the advanced, built-in security features and compliance certifications of leading cloud providers.
- Unlocking Data-Driven Insights: Use real-time analytics to make smarter, faster decisions that drive measurable business outcomes.

But this isn't just a technical manual or a checklist for IT teams. It's a strategic playbook for business leaders, technology decision-makers, and consultants alike. Whether you're an executive struggling to align your operations with modern market demands, a CIO navigating the complexities of cloud migration, or a consultant guiding clients through digital transformation, this book offers the insights, strategies, and practical tools you need to:

- Assess the true cost and impact of your technical debt.
- Build a compelling business case for cloud transformation.
- Develop a robust, comprehensive cloud migration strategy tailored to your organization.
- Choose the right cloud solutions to align with both your current operations and future goals.
- Mitigate risks and overcome common challenges during your transition.
- Maximize the long-term value of your cloud investment.

Throughout this journey, we'll move beyond the buzzwords and hype surrounding the cloud to focus on what truly matters: achieving your business objectives and creating a foundation for sustainable growth. This is a roadmap for leaders who know that transformation isn't just about technology; it's about unlocking potential, overcoming barriers, and redefining what's possible.

A Call to Action

As you turn these pages, imagine the possibilities: streamlined operations, empowered teams, and a future where your organization isn't held back by its systems but propelled by them. This is the promise of cloud transformation - a chance to break free from the limitations of the past and build a future of agility, innovation, and resilience.

The path to transformation won't be without challenges, but with the right strategies and tools, it's a journey worth taking. Let this book be your guide, and together, we'll unlock the true power of the cloud to drive your organization forward.

Let's get started. Your future awaits.

Chapter 1: COVID Breaks Everything - A Case Study of Cloud Strategies

The COVID-19 pandemic was a black swan event that exposed vulnerabilities in even the most advanced companies, forcing them to rethink their strategies and adapt to a new normal. The retail sector, home to giants like Walmart and Target, experienced these disruptions acutely. These companies, once considered leaders in logistics, supply chain management, and customer experience, faced unprecedented challenges that shook the foundations of their business models. For years, these companies had meticulously fine-tuned their operations, relying on just-in-time inventory models, seamless in-store experiences, and impulse-driven shopping behaviors. Yet, when the pandemic hit, these foundational strengths suddenly became liabilities.

The pandemic's impact was swift and severe. In-store shopping, the cornerstone of Walmart's and Target's revenue, evaporated overnight. Customers, once drawn by the convenience of a one-stop-shop experience, were forced to stay home. Impulse purchases, a significant driver of profitability, dropped to near-zero. The result was a severe blow to in-store revenue, forcing these companies to scramble for ways to engage customers who were no longer walking through their doors.

Simultaneously, demand for essential goods like cleaning supplies, groceries, and home office equipment surged. Shelves emptied faster than supply chains could refill them, exposing the fragility of just-in-time inventory models. These models, optimized for efficiency and cost savings, were not built to handle the sudden and universal spike in demand. The toilet paper and paper towel shortage became a symbol of this deeper crisis, highlighting the inability of traditional supply chains to handle the "everyone buying the same thing at the same time" phenomenon.

Walmart and Target had to pivot quickly, not just to survive but to keep up with the rapidly changing needs of their customers. They had to reimagine how to engage customers online, re-engineer their supply chains, and fundamentally shift their focus from efficiency to resilience. This transformation was enabled by the cloud, which provided the scalability, flexibility, and agility needed to adapt to the new reality.

COVID Changes the B2C Model Overnight

One of the most pressing challenges was enabling a safe and efficient way for customers to shop. Social distancing requirements made in-store shopping untenable for many, accelerating the adoption of curbside pickup. Walmart and Target had to scale this service rapidly, which required a robust digital infrastructure capable of real-time inventory visibility, seamless order management, and instant communication with customers. Cloud technology became the backbone of this transformation.

Within weeks, Walmart and Target re-engineered their supply chains and fulfillment processes. Stores were converted into mini-distribution centers, where employees, equipped with cloud-connected devices, picked and packed orders for customers waiting in their cars. Real-time updates allowed customers to track their orders, while mobile apps facilitated contactless payment and communication. What would typically have taken years to implement became a reality in a matter of weeks, thanks to the flexibility of cloud-based platforms.

The Cloud as the Enabler of Overnight Transformation

What set Walmart and Target apart during the pandemic wasn't just their size or resources; it was their ability to leverage cloud technology to adapt with unprecedented speed. Traditional IT systems, often characterized by silos and inflexibility, would have crumbled under the pressure of such rapid change. The cloud, however, offered:

Elasticity: As e-commerce traffic surged, cloud platforms scaled automatically to handle the load, ensuring seamless customer experiences.

Integration: By unifying data across physical stores, online platforms, and supply chains, the cloud provided a single source of truth for decision-making.

Innovation: Cloud-native development platforms enabled rapid deployment of new features, such as curbside pickup apps and personalized recommendation engines.

Resilience: With real-time analytics and predictive capabilities, these companies could respond to disruptions faster than ever before.

The transformation of Walmart and Target during the pandemic wasn't just a response to a one-time crisis; it set a new standard for business agility. These companies emerged stronger, with digital capabilities that will continue to drive growth long after the pandemic. For businesses across industries, their journey offers a powerful lesson: In a world where change is constant, the cloud isn't just a tool; it's the foundation for resilience, innovation, and success.

Deep Dive into the Challenges and Solutions

To truly understand the scale of the transformation, let's take a closer look at the specific challenges faced by Walmart and Target, and how they leveraged cloud technology to overcome them.

The Immediate Shock: The Loss of In-Store Revenue

Before the pandemic, the majority of Walmart's and Target's revenues came from in-store shopping. Customers didn't just visit these stores to grab essentials; they were drawn by the convenience of a one-stop-shop experience, often leaving with far more than they originally intended. This dynamic drove substantial revenue, but it relied heavily on physical foot traffic. When the world locked down, that traffic disappeared almost instantly. In-store impulse purchases - the hallmark of these businesses - dropped to near-zero. A customer who might have gone in for paper towels and left with patio furniture and a barbecue grill was now ordering a single item online. The result was a severe blow to profitability and a scramble to reimagine how to engage customers who were no longer walking through the doors.

The Supply Chain Crisis: From Efficiency to Resilience

The pandemic also exposed a harsh truth about global supply chains: they were optimized for efficiency, not resilience. Just-in-time inventory models, which minimized costs by keeping stock levels low, left companies vulnerable to disruptions. Walmart and Target, with their complex global supply chains, had to pivot rapidly to secure critical inventory.

Using cloud-based supply chain platforms, they gained real-time visibility into their networks, identifying bottlenecks and reallocating resources on the fly. Advanced forecasting models, powered by cloud analytics, allowed them to predict demand spikes and adjust procurement strategies accordingly. For example, when demand for cleaning supplies surged, these companies used predictive analytics to prioritize shipments from suppliers and reroute inventory to high-need regions. The cloud enabled these decisions to happen in hours rather than weeks—a crucial advantage in a rapidly evolving crisis.

The Shift to Digital and Curbside Pickup

Among the most pressing challenges was enabling a safe and efficient way for customers to shop. Social distancing requirements made in-store shopping untenable for many, accelerating the adoption of curbside pickup—a service that, before 2020, had been more of a value-add than a core capability. For Walmart and Target, scaling curbside pickup wasn't just a matter of operational logistics; it required a digital infrastructure capable of real-time inventory visibility, seamless order management, and instant communication with customers. Cloud technology became the backbone of this transformation.

The Role of Cloud Technology

The cloud played a crucial role in enabling Walmart and Target to overcome these challenges. It provided the scalability, flexibility, and agility needed to adapt to the new reality. Here's how:

Scalability: Cloud platforms scaled automatically to handle the unprecedented surge in e-commerce traffic, ensuring seamless customer experiences.

Integration: The cloud unified data across physical stores, online platforms, and supply chains, providing a single source of truth for decision-making.

Innovation: Cloud-native development platforms enabled rapid deployment of new features, such as curbside pickup apps and personalized recommendation engines.

Resilience: With real-time analytics and predictive capabilities, these companies could respond to disruptions faster than ever before.

The New Standard for Business Agility

The transformation of Walmart and Target during the pandemic wasn't just a response to a one-time crisis; it set a new standard for business agility. These companies emerged stronger, with digital capabilities that will continue to drive growth long after the pandemic. For businesses across industries, their journey offers a powerful lesson: In a world where change is constant, the cloud isn't just a tool; it's the foundation for resilience, innovation, and success.

Conclusion

The COVID-19 pandemic forced businesses to confront unprecedented challenges, and the retail industry was no exception. Walmart and

Target's ability to adapt quickly and effectively was a testament to their strategic embrace of cloud technology. The cloud enabled them to reimagine their operations, from re-engineering their supply chains to enhancing their e-commerce platforms. This transformation not only allowed them to weather the storm but also positioned them for continued growth in the post-pandemic world. For businesses looking to build resilience and thrive in an unpredictable world, the lesson is clear: the cloud isn't just an option; it's a necessity.

The COVID-19 pandemic underscored the critical role of technology in business resilience and agility. Walmart and Target's ability to leverage the cloud for rapid adaptation highlights a crucial aspect of navigating such disruptions: minimizing technical debt. In the following chapter, we'll delve deeper into the concept of technical debt and how it can hinder a company's ability to innovate and respond effectively to unforeseen challenges.

Chapter 2: Understanding Technical Debt

Introduction

In the previous chapter, we explored how the COVID-19 pandemic forced businesses like Walmart and Target to confront unprecedented challenges and adapt quickly to survive. Their ability to leverage technology, particularly cloud computing, played a crucial role in successfully navigating these disruptions. They were able to quickly adapt to changing consumer demands and supply chain disruptions because their systems were agile and innovative. However, the ability to adapt and innovate isn't solely determined by the availability of technology but also by the presence or absence of technical debt. In this chapter, we'll delve deeper into the concept of technical debt and how it can hinder a company's ability to innovate and respond effectively to unforeseen challenges, as well as how to manage it, particularly in the context of cloud transformation.

What is Technical Debt?

Technical debt is a metaphor that describes the implied cost of rework caused by choosing an easy solution now instead of using a better approach that would take longer. It's the accumulation of shortcuts, compromises, and suboptimal choices made during the software development process. Like financial debt, technical debt can accumulate interest, making it more costly to address in the future. Essentially, it's the price you pay for prioritizing speed over quality in software development.

Forms of Technical Debt

Technical debt isn't just about messy code. It can manifest in various forms, each with its own implications:

Code Debt: This is the most common type of technical debt, resulting from poorly written, undocumented, or untested code. It can lead to increased maintenance costs, decreased performance, and difficulty in adding new features.

Design Debt: This arises from inadequate or outdated design choices. It can make it challenging to scale the system, integrate with other systems, or adapt to changing requirements.

Architecture Debt: This stems from flawed architectural decisions that can have significant consequences for a system's performance, scalability, and maintainability. It can make it difficult to modify or extend the system's functionality to meet future needs or accommodate growth, as well as limit the ability to adopt new technologies or integrate with other systems.

Infrastructure Debt: This is the result of outdated or poorly maintained infrastructure, which can lead to performance issues, security vulnerabilities, and increased operational costs.

Test Debt: Insufficient or ineffective testing can lead to undetected bugs and errors, resulting in costly rework and potential damage to the company's reputation.

Documentation Debt: Lack of clear and up-to-date documentation can hinder understanding and maintenance of the system, leading to errors and delays.

Technical Debt: Causes and Consequences

Technical debt is a significant challenge for many organizations, often resulting from several contributing factors. While it can enable quick wins or faster time-to-market, it accumulates long-term costs that can hinder innovation and operational efficiency. Below are the primary causes of technical debt, expanded with real-world context:

1. Business Pressure

In a competitive market, businesses often prioritize speed over quality to deliver products, features, or updates. For example, startups under pressure to secure funding might release a minimum viable product (MVP) that lacks robust architecture, leading to rework later.

Example: A company releasing a rushed e-commerce platform to capitalize on a seasonal shopping trend may forgo scalability considerations, leading to crashes under high traffic during peak periods.

2. Lack of Skills or Experience

Inexperienced or insufficiently skilled developers may write suboptimal code, unaware of best practices or architectural patterns. Organizations that skimp on training or hire underqualified staff risk introducing avoidable technical issues.

Example: A small firm developing an internal tool without senior oversight might create a system that works initially but cannot scale or integrate with other platforms, requiring expensive rewrites later.

3. Poor Communication

Inadequate communication between teams or departments can lead to inconsistent goals, misunderstood requirements, and misaligned implementations. These gaps often result in duplicated efforts or functionality gaps.

Example: A development team implementing a customer feedback feature might misinterpret marketing's requirements, creating a tool that doesn't align with user needs and requires iterative fixes.

4. Lack of Standards or Guidelines

The absence of coding standards or clear development guidelines can result in inconsistent, difficult-to-maintain codebases. Without enforcing best practices, teams may create technical silos, hindering collaboration and scalability.

Example: A global enterprise with distributed teams using different frameworks and coding styles might find their systems incompatible, slowing down future enhancements or integrations.

5. Deferred Refactoring

Organizations often deprioritize refactoring - improving the design and structure of existing code - because it doesn't deliver immediate visible value. Over time, this neglect leads to a codebase that is harder to maintain and extend.

Example: A legacy ERP system patched repeatedly over years without refactoring can become so complex that even minor changes risk breaking critical functionality, delaying modernization efforts.

This standalone narrative provides detailed insights into the root causes of technical debt and their potential consequences, offering clarity for discussions or documentation. Let me know if you'd like to add or adjust anything further!

Impact of Technical Debt

Technical debt can significantly impact businesses, affecting their bottom line, agility, and overall health. Let's take a closer look at the key consequences:

Increased Costs

Addressing technical debt can be expensive, requiring significant time and resources for rework and refactoring. Imagine a company with a large, monolithic application built on outdated technology. Modernizing this application could involve rewriting significant portions of the codebase, migrating to a new platform, and retraining staff. This can be a costly and time-consuming process, especially compared to the initial "quick and easy" solution that led to the technical debt.

Reduced Agility

Technical debt can make it difficult to adapt to changing market conditions or customer needs, hindering innovation and competitiveness. Think of a company operating in the fast-moving world of e-commerce. If their systems are burdened by technical debt, they may struggle to implement new features, integrate with new platforms, or respond to emerging trends. This can lead to lost opportunities and a decline in market share.

Decreased Performance

Poorly written or poorly designed code can lead to performance issues, impacting user experience and productivity. For example, a slow-loading website can frustrate customers and drive them away. Similarly, a buggy application can hamper employee productivity and lead to costly errors. Addressing these performance issues often requires significant effort to identify and fix the underlying technical debt.

Increased Risk

Technical debt can increase the risk of security vulnerabilities, data breaches, and system failures. Outdated systems and insecure coding practices can leave companies exposed to cyberattacks. A data breach can damage a company's reputation, lead to financial losses, and erode customer trust. Addressing these security risks often involves upgrading systems, patching vulnerabilities, and implementing security best practices, all of which can be costly and time-consuming.

Reduced Morale

Working with a codebase burdened by technical debt can be frustrating and demotivating for developers. Imagine spending hours trying to understand and modify a tangled mess of undocumented code. This can lead to burnout, decreased productivity, and even attrition. Creating a healthy and sustainable development environment requires investing in clean code, clear documentation, and supportive tools, which can help reduce technical debt and boost developer morale.

In conclusion, technical debt can have a significant impact on businesses, affecting their financial performance, agility, and overall health. By understanding the consequences of technical debt and taking proactive steps to manage it, companies can minimize its negative impact and position themselves for success in today's rapidly changing business environment.

Recognizing the Symptoms of Technical Debt

Technical debt, like a hidden leak in your home's foundation, may not always be immediately visible. However, there are telltale signs that it's starting to weigh your systems down and hinder your business's agility and efficiency.

Frequent System Downtime

Legacy infrastructure and outdated systems may struggle to keep up with modern workloads, leading to frequent crashes and downtime. Imagine a retail company relying on an old server infrastructure to handle online orders during a peak shopping season. As traffic surges, the outdated system buckles under the pressure, resulting in website crashes and frustrated customers who can't complete their purchases. This not only leads to immediate revenue loss but also damages the company's reputation and customer trust.

Inefficient Integrations

In today's interconnected world, businesses rely on various software applications and cloud services. Difficulty connecting legacy systems with modern cloud platforms or SaaS tools can be a major warning sign of technical debt. For instance, a company might want to integrate its customer relationship management (CRM) system with a new marketing automation platform. However, if the CRM system is built on outdated technology, creating a seamless integration can be complex, time-consuming, and costly. This can hinder the company's ability to leverage the full potential of both systems and streamline its operations.

Developer Bottlenecks

If your developers are spending more time debugging and maintaining old code than innovating and building new features, it's a sign that technical debt is slowing them down. Imagine a software development team tasked with creating a new mobile app for a banking client. Instead of focusing on designing innovative features and user-friendly interfaces, they're bogged down trying to decipher and fix issues in a tangled mess of legacy code. This not only delays the project but also stifles creativity and reduces the team's overall productivity.

Resistance to Change

Leadership apprehension about the costs and risks of modernizing systems can be a symptom of technical debt. The longer technical debt is left unaddressed, the more daunting the task of fixing it becomes. This can lead to a culture of resistance to change, where even minor upgrades or updates are met with apprehension. For example, a company might be reluctant to migrate its on-premises infrastructure to the cloud due to concerns about the initial investment and the perceived complexity of the migration process. However, this resistance can prevent the company from realizing the long-term benefits of cloud computing, such as scalability, cost-efficiency, and enhanced agility.

By recognizing these symptoms, businesses can take proactive steps to address technical debt and prevent it from hindering their growth and innovation. In the following sections, we'll explore strategies for managing technical debt and leveraging the cloud to modernize systems and enhance business agility.

Managing Technical Debt

Effectively managing technical debt is crucial for maintaining a healthy and sustainable software development process. Here are some strategies:

- Acknowledge and Track: The first step is to acknowledge the existence of technical debt and track it using appropriate metrics and tools. This will help you understand the scope of the problem and prioritize your efforts.
- Prioritize: Not all technical debt is created equal. Prioritize addressing the most critical issues that have the biggest impact on the business. Focus on the areas where the technical debt is causing the most pain, such as frequent system outages, performance bottlenecks, or security vulnerabilities.
- Refactor Regularly: Incorporate regular code refactoring into the development process to prevent the accumulation of technical debt. Refactoring is like cleaning up your workspace to make it more efficient and organized.
- Invest in Automation: Automate testing and other repetitive tasks to reduce errors and free up developers' time for more strategic work. Automation can help you catch bugs early on and reduce the amount of manual effort required to maintain your systems.
- Establish Clear Standards: Define and enforce coding standards and guidelines to ensure consistency and maintainability of the codebase. Having clear standards makes it easier for developers to understand and work with the code, reducing the likelihood of introducing new technical debt.
- Foster Communication: Encourage open communication and collaboration among team members to prevent misunderstandings and inconsistencies. Good communication can help prevent technical debt from arising in the first place.

Invest in Training: Provide developers with the necessary training and resources to improve their skills and knowledge. This will help them make better decisions and write better code, reducing the accumulation of technical debt.

The Role of the Cloud in Addressing Technical Debt

Cloud computing can play a significant role in managing and reducing technical debt. It offers a range of capabilities that enable businesses to modernize their systems, improve agility, and reduce costs.

Scalability

Cloud platforms can scale resources up or down as needed, reducing the risk of infrastructure debt. This means you can easily adjust your computing capacity to meet changing demands without having to worry about managing and maintaining your own hardware. For example, a retailer experiencing a surge in online orders during a holiday season can quickly scale up its cloud resources to handle the increased traffic and prevent website crashes. This scalability helps avoid the costs and complexities associated with managing and upgrading physical infrastructure.

Use Case: A streaming service experiences a sudden surge in subscribers due to a popular new release. With cloud scalability, they can automatically provision more servers to handle the increased demand, ensuring uninterrupted streaming for their customers.

Cost Efficiency

Cloud services offer a shift from CapEx-heavy on-premises systems to OpEx-focused cloud models. This can free up resources that can be used to address existing technical debt or invest in new technologies. Instead of investing large sums upfront for hardware and software licenses, businesses can pay for cloud services on a subscription basis, reducing upfront costs and allowing for greater financial flexibility. This can be particularly beneficial for startups and small businesses with limited resources.

Use Case: A startup company needs to build a data analytics platform but has limited capital. By using cloud services, they can avoid the upfront costs of purchasing and maintaining servers, allowing them to allocate more resources to developing their core product.

Streamlined Integrations

Modern APIs and middleware facilitate seamless connections between legacy and modern systems. This can help you break down data silos and integrate your systems more effectively, reducing the complexity of your IT environment. For example, a company can use cloud-based integration services to connect its legacy accounting system with a modern e-commerce platform, enabling real-time data synchronization and improved business insights.

Use Case: A healthcare provider wants to integrate patient data from various legacy systems into a unified electronic health record (EHR) system. Cloud-based integration services can help them connect these disparate systems, enabling seamless data sharing and improved patient care.

Enhanced Agility

Cloud platforms accelerate innovation and deployment cycles, reducing the burden of technical debt. This means you can respond more quickly to changing market conditions and customer needs, giving you a competitive edge. For example, a software company can leverage cloud-based development tools and continuous integration/continuous delivery (CI/CD) pipelines to rapidly develop, test, and deploy new features and updates to its applications. This agility helps businesses stay ahead of the curve and adapt to evolving customer demands.

Use Case: A financial institution wants to launch a new mobile banking app with innovative features. By using cloud-based development platforms and CI/CD pipelines, they can accelerate the development process and release the app to market faster, gaining a competitive advantage.

Key Strategies to Reduce Technical Debt

Here are some key strategies to reduce technical debt, particularly in the context of cloud transformation:

Refactor vs. Replace

Decide whether to refactor existing systems or replace them entirely based on cost-benefit analysis. Refactoring can be a good option for improving the structure and maintainability of existing code, while replacing a system entirely might be necessary if it's too outdated or complex to refactor. For example, a company might choose to refactor a critical application that is still functional but has accumulated significant technical debt over time. On the other hand, a legacy system that is no longer supported or cannot meet current business needs might be a candidate for replacement with a modern cloud-based solution.

Use Case: A company has a core application written in an older programming language. Refactoring it to a modern language could improve performance and maintainability. However, if the application is heavily reliant on outdated libraries or frameworks, replacing it with a new cloud-native application might be a more viable option.

Adopt Modular Architectures

Implement microservices and containerization to replace monolithic systems gradually. This can make your systems more flexible and easier to maintain, as well as reduce the impact of changes to individual components. By breaking down applications into smaller, independent services, businesses can update or replace specific components without affecting the entire system. This approach allows for greater agility and reduces the risk of introducing new technical debt.

Use Case: An e-commerce company wants to improve the scalability and resilience of its online platform. By adopting a microservices architecture, they can break down the platform into smaller, independent services, such as product catalog, shopping cart, and payment processing. This allows them to scale individual services as needed and update or replace components without disrupting the entire platform.

Invest in Training

Equip teams with the skills needed to transition and maintain modern solutions. This will help them make better decisions and write better code, reducing the accumulation of technical debt. Investing in training and development programs can help developers stay up-to-date with the latest technologies and best practices, enabling them to build and maintain high-quality software.

Use Case: A company is migrating its infrastructure to the cloud. Investing in cloud training for its IT staff will equip them with the necessary skills to manage and maintain the new cloud environment, reducing the risk of errors and technical debt.

Plan for Long-Term Value

Align technical debt reduction initiatives with strategic business goals. This will ensure that your efforts are focused on the areas that will have the biggest impact on your business. For example, if a company's strategic goal is to improve customer experience, then technical debt reduction efforts should focus on modernizing systems that directly impact customer interactions, such as the company's website or mobile app.

Use Case: A bank wants to improve its digital banking services to attract and retain customers. Aligning its technical debt reduction efforts with this goal, the bank focuses on modernizing its online banking platform and mobile app, ensuring a seamless and user-friendly experience for its customers.

Collaborate Cross-Functionally

Engage stakeholders across departments to ensure alignment on priorities and outcomes. This will help you build consensus and get buy-in for your technical debt reduction initiatives. Addressing technical debt often requires collaboration between IT, business units, and leadership. By involving stakeholders from different departments, businesses can ensure that technical debt reduction efforts are aligned with overall business objectives and priorities.

Use Case: A manufacturing company wants to implement a cloud-based supply chain management system. By collaborating with stakeholders from procurement, logistics, and production, the company can ensure that the new system meets the needs of all departments and integrates seamlessly with existing processes.

Framing the Business Case for Addressing Technical Debt

When proposing initiatives to address technical debt, it's crucial to present a compelling business case that resonates with executive decision-makers such as the CFO and the VP of Operations. These stakeholders are primarily concerned with financial outcomes, operational efficiency, and strategic alignment. Here's an expanded approach to framing the business case, complete with an example scenario:

1. ROI of Addressing Technical Debt

Highlight the tangible benefits of addressing technical debt by focusing on metrics that matter to the CFO and VP of Operations:

Improved Efficiency: Demonstrate how resolving technical debt can streamline operations, reduce system downtime, and lower maintenance costs.

Cost Savings: Illustrate the long-term financial benefits of migrating to modern, cloud-based solutions, such as reduced infrastructure costs and minimized risk of unexpected outages.

Faster Innovation Cycles: Emphasize how addressing technical debt will free up resources currently tied to "keeping the lights on," enabling the organization to focus on strategic initiatives.

Example:
A manufacturing company is running on a legacy ERP system that requires constant manual intervention to process orders and synchronize inventory across its facilities. These inefficiencies cost the company $1 million annually in overtime labor and lost productivity. By addressing technical debt through an integrated, cloud-based ERP solution, the company can eliminate manual processes, save $800,000 annually, and achieve a 20% increase in operational efficiency.

2. Customer-Centric Value

Focus on the direct impact technical debt has on customer satisfaction and retention:

Enhanced Customer Experience: Explain how outdated systems lead to slower response times, poor user interfaces, and missed opportunities to delight customers.

Quicker Time-to-Market: Show how reducing technical debt enables faster delivery of new features and capabilities that customers demand.

Example:
In the current system, a retail company takes three weeks to update its e-commerce platform with new promotions due to hardcoded dependencies and outdated workflows. During the holiday season, this delay results in $500,000 in missed sales opportunities. By modernizing the platform, the company can reduce update time to two days, enabling real-time promotions that drive customer engagement and boost seasonal revenue.

3. Future-Proofing the Organization

Position the reduction of technical debt as a strategic investment that prepares the organization for future opportunities and challenges:

- **Adaptability to Emerging Technologies**: Demonstrate how reducing technical debt creates the agility needed to adopt AI, machine learning, IoT, and other emerging technologies.
- **Scalability:** Highlight how modern systems allow the business to scale operations seamlessly, supporting growth without exponential increases in costs.
- **Mitigation of Risks:** Stress that addressing technical debt reduces vulnerabilities to cyberattacks, compliance failures, and system failures.

Example:

A logistics company's legacy system cannot integrate with modern supply chain analytics tools. This prevents real-time tracking and predictive optimization, leaving the company unable to compete with industry leaders leveraging advanced analytics. By addressing technical debt and implementing an API-friendly cloud platform, the company gains real-time insights, improves delivery accuracy by 15%, and achieves a 10% reduction in logistics costs.

Scenario: Presenting to the CFO and VP of Operations

Background:

The organization is a regional retail chain experiencing significant inefficiencies due to its outdated inventory management system. Frequent stockouts, overstock issues, and manual data entry processes result in $2 million in annual losses. IT leadership proposes a project to modernize the inventory system, addressing technical debt and implementing a cloud-based solution.

Presentation Approach:

Introduction:

Start with a high-level overview: "Our current inventory system is costing us $2 million annually due to inefficiencies, and we're at risk of losing customers to competitors with more agile systems. By addressing technical debt and modernizing this system, we can save $1.5 million annually and position ourselves for long-term growth."

ROI Breakdown:

Quantify the financial impact: "By eliminating manual processes, we reduce labor costs by $750,000 annually. Improved inventory accuracy minimizes stockouts and overstocks, saving an additional $500,000. A cloud-based system reduces IT infrastructure costs by $250,000."

Customer-Centric Benefits:

Emphasize customer experience: "Our customers expect accurate inventory availability and fast fulfillment. A modernized system will enable real-time updates, ensuring we can meet customer demands and increase satisfaction."

Future-Proofing:

Address strategic growth: "This initiative positions us to leverage predictive analytics and AI tools for demand forecasting, ensuring we remain competitive in the market. It also enables seamless scalability as we expand to new regions."

Risk Mitigation:

Highlight risks of inaction: "If we don't address this now, our outdated system could fail during peak demand, leading to even greater losses. Competitors are already investing in similar upgrades, which puts us at a disadvantage."

Implementation Plan:

Provide a phased approach: "The project will be executed in three phases over 12 months, minimizing disruptions. Each phase will deliver incremental improvements, with ROI visible within the first six months."

Closing Argument:

Summarize the value: "This is more than a technology upgrade; it's a strategic investment that will reduce costs, enhance customer satisfaction, and future-proof our operations. Addressing this technical debt is essential to our continued success."

Key Takeaways for Decision-Makers

For the CFO: The financial argument centers on measurable cost savings, ROI, and risk mitigation, making it clear that addressing technical debt is a fiscally responsible choice.

For the VP of Operations: The operational benefits—such as efficiency gains, improved workflows, and scalability—directly align with their priorities of streamlining processes and supporting growth.

By aligning the business case with the priorities of each stakeholder, you create a compelling argument for addressing technical debt that's difficult to ignore. Let me know if you'd like to refine this further or add more examples!

Conclusion

Technical debt is an unavoidable reality for businesses with IT systems, especially those involving code development. It's not necessarily a sign of poor practices, but rather a consequence of the constant need to balance speed, cost, and quality. By understanding the causes and consequences of technical debt, and by implementing effective management strategies, businesses can minimize its negative impact and maintain a healthy and sustainable development lifecycle.

The cloud offers a powerful toolkit for addressing technical debt, providing scalability to adapt to changing demands, cost-efficiency to free up resources, streamlined integrations to connect disparate systems, and enhanced agility to accelerate innovation. Proactive technical debt management, empowered by the cloud, allows businesses to improve their agility, reduce costs, and position themselves for success in today's dynamic business environment.

While understanding technical debt is crucial, it's equally important to recognize the broader context in which it exists. Technical debt is often a symptom of a deeper need for organizational transformation. In the next chapter, we'll explore why transformation is essential for businesses to thrive in the digital age and how to effectively make the case for change within your organization.

Chapter 3: Building a Cloud-First Vision

Moving to the cloud isn't just about lifting and shifting existing applications; it's like upgrading from a horse-drawn carriage to a self-driving car - you don't just strap a rocket to the old buggy and hope for the best. It's about fundamentally reimagining how your business operates and leveraging technology to drive innovation, agility, and growth. This chapter explores what a truly modernized business looks like in the cloud era, how cloud solutions can act as enablers, and the crucial role of leadership in driving this transformation.

The Hidden IT Company Inside Every Business

Whether you're a retailer, a manufacturer of paper plates, or running an online dating service, you've unwittingly become the proud owner of an IT department. Yes, that's right - you're in the computer business along with your main gig. For our paper plate manufacturer, this means you're not just dealing with pulp and packaging but also servers, networks and maybe even a room full of blinking lights that only Jeff from IT dares to enter.

You're spending precious capital on servers, equipment, HVAC systems to keep that server room chilly enough for penguins and hiring expensive IT personnel to manage and operate it all. Then you have to concern yourself with security breaches, disaster recovery planning (hope those backup tapes are stored safely!), server and equipment maintenance... Whew - are you tired yet? It's like juggling flaming torches while riding a unicycle - not impossible, but far from ideal.

Bridging the Gap: From IT Burden to Strategic Advantage

For most businesses, managing IT in-house has become an expensive, resource-draining necessity rather than a strategic advantage. The question is no longer whether technology plays a role in your business

- it's whether your business is leveraging technology in the smartest, most scalable way.

This is where cloud maturity comes into play. Companies that embrace cloud solutions shift from wrestling with IT infrastructure to harnessing technology for innovation and growth. But not all cloud journeys are created equal. Some businesses are still dipping their toes in the water, while others are fully leveraging cloud-native capabilities to drive competitive advantage.

So, where does your company fall on the cloud maturity curve? Let's take a closer look at how to benchmark your cloud adoption and uncover where the real opportunities for transformation lie.

Benchmarking Cloud Maturity

To ensure a successful cloud transformation, businesses must understand their starting point. Cloud readiness assessments, such as the Cloud Adoption Framework by major providers (e.g., Microsoft, AWS, Google Cloud), can help organizations gauge their preparedness. These assessments evaluate factors like infrastructure, skill sets, governance, and security readiness, offering a roadmap to maturity. For example, a manufacturing company can use these tools to prioritize migrating legacy ERP systems before scaling to advanced analytics solutions.

Understanding where your organization stands in terms of cloud maturity allows for targeted investments and strategic planning. It's not just about identifying gaps but recognizing strengths that can be leveraged. For instance, a business with robust data management practices might accelerate its transition by focusing on scalable analytics platforms. On the other hand, companies lacking governance policies may need to establish foundational processes before tackling complex migrations. Benchmarking ensures that cloud adoption is not only feasible but sustainable over time.

Decision-Making in Action

Consider a logistics company that implemented a cloud-based platform for real-time data aggregation across its fleet. By analyzing data from IoT sensors, the company optimized routes, reduced fuel consumption by 15%, and improved delivery timelines. This transformation showcases how cloud-enabled insights can drive significant operational efficiencies and cost savings.

This example highlights the transformational impact of data as a strategic asset. When organizations harness the power of real-time data analytics, they move from reactive decision-making to proactive, insight-driven strategies. A retailer could similarly analyze customer preferences in real time to personalize promotions, driving sales and improving customer loyalty. The shift to leveraging data is not a mere upgrade; it redefines how businesses operate, ensuring they remain competitive in a rapidly evolving landscape.

What Does a Modernized Business Look Like?

Imagine a business where:

- Data is a strategic asset: Data isn't just collected; it's seamlessly analyzed and used to make informed decisions in real-time. Picture a retailer that personalizes customer experiences based on individual preferences and purchase history - like a barista who starts making your favorite drink the moment you walk in. Or a manufacturer optimizing production based on real-time sensor data from the factory floor, preventing breakdowns before they happen.
- Processes are automated and efficient: Manual tasks are minimized, workflows are streamlined, and employees are empowered to focus on high-value activities. Think of a finance department that automates invoice processing and reconciliation - no more chasing paper trails or deciphering cryptic handwriting. Or a human resources team using AI-powered tools to streamline recruitment and onboarding, so new hires feel like part of the family from day one.

- Collaboration is seamless: Teams work together effortlessly, regardless of location or device. Imagine a global team collaborating on a product launch using shared documents and video conferencing, brainstorming as if they were all in the same room - even if one of them is in pajamas halfway around the world.
- Innovation is accelerated: New products and services are developed and deployed rapidly, responding to market demands and customer needs. Envision a software company continuously delivering new features and updates to its customers, or a healthcare provider using cloud-based analytics to develop personalized treatments faster than you can say "open wide."
- Security is paramount: Data and systems are protected by robust security measures, ensuring compliance and minimizing risks. Think of a financial institution leveraging cloud-based security tools to detect and prevent fraud, staying one step ahead of cyber threats.

This is the vision of a modernized business - empowered by the cloud to achieve greater efficiency, agility, and innovation. It's like swapping out your old flip phone for the latest smartphone; you're not just making calls - you're accessing a world of possibilities.

Cloud as an Enabler

Cloud solutions are more than just infrastructure; they are catalysts for business transformation. They provide organizations with the scalability, agility, and innovation needed to thrive in today's fast-paced digital economy - all while optimizing costs and improving collaboration.

Scalability: Expanding Without the Growing Pains

Imagine an online retailer preparing for the holiday shopping season. Traditionally, they'd have to invest heavily in physical servers just to handle a temporary traffic surge—a bit like buying a fleet of buses for one rush hour and then letting them collect dust. With cloud-based infrastructure, they can dynamically scale up computing resources when demand spikes and scale back down afterward. This ensures smooth customer experiences without overpaying for unused capacity - a game-changer for businesses with seasonal fluctuations.

Agility: Moving at the Speed of Change

In a world where trends shift overnight, speed matters. Take a mobile app developer who notices a competitor launching a breakthrough feature. Instead of being bogged down by lengthy development cycles, they leverage cloud-native tools to rapidly prototype, test, and deploy updates in days rather than months. It's like swapping a slow-turning cruise ship for an agile speedboat - you can pivot instantly without losing momentum.

Innovation: Unlocking New Possibilities

The cloud isn't just about running existing systems more efficiently; it's about creating entirely new opportunities. A healthcare startup, for example, might use cloud-based AI and machine learning to analyze patient data and predict potential health risks before they escalate. Without cloud computing, accessing such computational power would be prohibitively expensive. Now, even small players can leverage cutting-edge technology once reserved for industry giants - like having a supercomputer at their fingertips.

Cost Optimization: Doing More with Less

For many organizations, the cloud is the difference between strategic investment and IT overhead bloat. A small law firm moving its document management system to the cloud no longer has to maintain on-premises servers or worry about expensive hardware failures. Instead of constantly troubleshooting IT issues, their staff can focus on improving client services It's the difference between owning a car (with all its maintenance woes) and using a rideshare app - you get where you need to go without the headaches.

Collaboration: Breaking Down Silos

Workforces today are more distributed than ever. A global advertising agency with teams in New York, London, and Tokyo can collaborate seamlessly using cloud-based tools. Designers, copywriters, and clients all work on the same files in real-time, eliminating version confusion and streamlining feedback. It's like having a virtual office where everyone's just a swivel chair away, minus the awkward elevator rides.

All Cloud Strategies Must Be Driven by a Business Case

Before embarking on any cloud journey, organizations must establish a clear and compelling business case. This isn't just a formality—it's the guiding star that ensures cloud investments align with strategic goals, deliver measurable value, and avoid becoming a solution in search of a problem.

Take, for example, a regional grocery chain aiming to enhance its online shopping platform. Instead of adopting cloud technology for the sake of modernization, leadership first identified key business drivers: reducing checkout times, increasing order accuracy, and boosting online sales. The business case projected that a cloud-based solution would cut checkout times by 30% and increase digital revenue by 20% within a year—concrete metrics that justified the investment and ensured accountability.

Similarly, a manufacturing company considering a cloud-based supply chain solution must first evaluate the expected return on investment (ROI). Will the move improve delivery timelines? Reduce operational costs? Enhance supplier collaboration? Without clear benchmarks, the project risks becoming a costly experiment rather than a strategic advantage.

A well-structured business case doesn't just secure funding—it defines success, aligns stakeholders, and provides a roadmap for execution. Cloud adoption works best when it's not just an IT initiative, but a business-driven transformation with tangible impact.

Avoiding Common Pitfalls

Cloud initiatives often fail due to poor alignment with business objectives. Without a clear business case, organizations risk budget overruns and low adoption rates. For instance, a retailer that adopted a cloud CRM without integrating it with existing systems faced low user engagement and missed sales opportunities. Establishing clear goals, such as "reduce customer response time by 20%," mitigates these risks and ensures measurable impact.

The foundation of a successful cloud strategy lies in understanding the "why." Why is the cloud necessary, and how does it serve the organization's broader goals? By answering these questions, businesses can craft strategies that resonate across departments. For example, a hospital implementing a cloud-based patient records system must align its goals with outcomes like improved patient care and compliance with healthcare regulations. A lack of alignment turns a promising initiative into an expensive experiment, emphasizing the need for clear, measurable objectives.

Tracking Success Metrics

Effective cloud strategies incorporate KPIs to measure success. Metrics such as "time-to-market improvement," "percentage reduction in IT overhead," and "uptime reliability" provide tangible evidence of progress. A financial services firm, for example, tracked a 25% reduction in server downtime post-cloud migration, highlighting operational resilience and cost savings.

Metrics act as the compass guiding cloud strategies. Beyond basic financial metrics, organizations should consider user adoption rates, customer satisfaction scores, and operational improvements. For example, a retail chain's ability to handle increased web traffic during peak shopping seasons is a direct measure of the cloud's scalability. These metrics not only justify investments but also build confidence among stakeholders, fostering long-term support for cloud initiatives.

Cloud Strategies Should Be Incremental Instead of Big Bang

While the idea of a sweeping cloud transformation is tempting, experience shows that an incremental approach delivers better results. A phased adoption allows organizations to test, learn, and adapt without overwhelming teams or risking major disruptions.

For example, instead of migrating everything at once, a financial services company might start with non-critical applications like payroll or HR systems. This approach provides valuable insights into migration challenges, user adoption, and performance benchmarks—lessons that can be applied to larger, mission-critical projects.

Use Case: Incremental Cloud Deployment

A mid-sized healthcare provider needed to modernize its IT infrastructure but wanted to avoid disruptions. Instead of an all-at-once transition, it started by migrating email and collaboration tools to the cloud. This small, controlled step built internal trust and confidence, setting the stage for the eventual migration of its electronic health records system.

By beginning with non-critical applications, the provider identified compatibility challenges early and resolved them before tackling mission-critical workloads. This phased approach not only minimized risks but also increased buy-in from stakeholders, reinforcing a culture of continuous improvement.

Incremental cloud adoption provides businesses with the breathing room to identify technical, cultural, or operational challenges before they escalate. A university, for instance, might migrate student records first, gathering insights before transitioning faculty systems. This method ensures progress while minimizing disruption.

Addressing Incremental Challenges

While a step-by-step approach is less disruptive, it's not without its own hurdles. Ensuring seamless integration between cloud and on-premises systems often requires robust middleware and API strategies. Additionally, organizations must address potential data silos and provide user training to facilitate adoption. A phased pilot program - with clear objectives for each stage - helps smooth transitions and builds confidence among employees.

Establishing feedback loops at every stage is crucial. These loops help teams identify gaps, measure success, and adjust strategies as needed. For example, a healthcare provider might introduce a cloud-based scheduling tool in a single department, gathering input from staff and patients before scaling across the organization. Addressing challenges in real time ensures that each small success lays the groundwork for larger, transformative projects.

The Role of Leadership: Inspiring Alignment and Trust

Cloud transformation isn't just an IT initiative - it's a fundamental cultural shift. Successful leaders must:

- Articulate a clear vision: Define business benefits and paint a compelling picture of the future. If the captain doesn't know the destination, the crew won't know which way to row.
- Foster a culture of change: Encourage experimentation, embrace new ways of working, and celebrate both successes and lessons from failures. Think of it as turning your company into a team of explorers, eager to discover new opportunities.
- Build trust and transparency: Address concerns proactively, provide clear communication, and ensure employees feel included in the process. Rumors fill a vacuum - better to replace them with facts and a shared sense of purpose.
- Invest in skills development: Equip employees with the necessary training and resources to thrive in a cloud-first environment. You wouldn't send a knight into battle without a sword - don't send your team into the cloud without the right tools.
- Lead by example: Demonstrate commitment to the cloud vision by actively using cloud-based tools and promoting their

benefits. If you're asking everyone to embrace change, show them you're first in line.

Leadership Styles and Cloud Success

Different leadership styles can significantly impact cloud transformation success. Transformational leaders inspire teams with a compelling vision, while participative leaders foster engagement and buy-in by involving employees in the decision-making process.

For instance, a healthcare CEO who was hands-on in implementing a cloud-based patient management system ensured cross-departmental alignment, accelerating adoption. Meanwhile, a manufacturing CIO might lead collaborative workshops to co-create cloud strategies with employees - turning potential resistance into enthusiastic support.

Cloud transformation leadership goes beyond setting direction - it's about embodying change. Transformational leaders energize teams with purpose and motivation, while participative leaders create a culture of inclusivity and shared ownership. By balancing both approaches based on organizational needs, leaders can ensure sustained success.

Case Study: Navigating Change

A global retail chain's CIO led a successful cloud transition by prioritizing transparency and communication. Weekly town halls addressed employee concerns, and a visible commitment to using the new tools inspired confidence. This proactive approach resulted in a 90% adoption rate within the first three months of rollout.

This case highlights a fundamental truth: leadership is the linchpin of any transformational initiative. Consistent communication, leading by example, and fostering trust are critical for ensuring employees embrace change. When leaders actively participate in the adoption of new tools and demonstrate commitment to transformation, it reduces apprehension and builds a cohesive, motivated workforce ready to tackle cloud migration challenges.

Effective leadership binds all elements of a successful cloud transformation. Without strong leadership, even the most sophisticated technology and well-defined strategies will falter. Leaders must inspire, motivate, and empower their teams to navigate challenges and realize the full potential of the cloud.

It's like assembling a symphony orchestra - you might have the finest musicians and instruments, but without a conductor to bring harmony, all you have is noise.

Emerging Trends and Future Considerations

While strong leadership is critical for today's cloud transformation, staying ahead means anticipating emerging trends that will define the future of cloud strategy. Businesses must continuously evolve, leveraging advancements such as multi-cloud strategies, edge computing, and artificial intelligence to remain competitive.

Multi-Cloud Strategies and Edge Computing

As organizations mature in their cloud adoption, many are shifting to multi-cloud strategies—using multiple cloud providers to diversify risk, enhance flexibility, and avoid vendor lock-in. Coupled with edge computing, which processes data closer to its source, these approaches enable real-time decision-making and improved performance.

For example, autonomous vehicle manufacturers rely on edge computing to process critical data locally, ensuring split-second reaction times. Similarly, a smart city initiative leveraging edge computing can dynamically optimize traffic lights based on real-time congestion, improving urban mobility. Together, these technologies enhance resilience and redefine what's possible in the cloud era.

AI's Role in Cloud Strategy

Artificial intelligence is revolutionizing cloud adoption by enhancing automation, optimizing workloads, and strengthening security. AI-driven tools can predict infrastructure needs, detect vulnerabilities, and streamline cloud migrations - unlocking new opportunities for innovation and operational efficiency.

Imagine a retail chain using AI to analyze shopping patterns, ensuring inventory levels match demand without overstocking. AI can also enhance cybersecurity by proactively identifying and mitigating threats before breaches occur. By embedding AI into cloud strategies, businesses position themselves at the forefront of technological advancement, ready to meet the challenges of tomorrow.

Conclusion

Leadership is the key to turning cloud adoption into business transformation. It's not just about moving to the cloud but about reshaping how an organization operates, innovates, and competes. A cloud-first vision requires a shift in mindset—one that sees the cloud not just as an IT upgrade but as an enabler of business transformation.

By envisioning a future where data is a strategic asset, processes are automated, and innovation accelerates, organizations can unlock the full potential of the cloud and thrive in the digital age. But technology alone is not enough. Strong leadership is essential to inspire alignment, foster a culture of change, and guide the organization toward a successful cloud-powered future.

It's time to pack up that old server room, give Jeff from IT a new mission, and set sail into the cloud. Trust me, the view from up here is spectacular.

Key Takeaways for Decision-Makers

✓ Conduct cloud readiness assessments to establish a strong foundation.

- ✓ Align cloud strategies with business objectives to ensure measurable ROI.
- ✓ Embrace incremental adoption for controlled, less disruptive transitions.
- ✓ Foster a leadership culture that champions transparency, skill development, and collaboration.
- ✓ Monitor emerging trends to future-proof cloud investments.

A Thought-Provoking Question

As businesses move into the cloud era, the question isn't just "What can the cloud do for us?" but rather "How can we reimagine our business to unlock its full potential?"

With strong leadership and a clear vision, organizations can achieve transformative outcomes—leaving legacy constraints behind and fully embracing the opportunities of a cloud-first future.

Chapter 4: Assessing Your Legacy Landscape

Cloud transformation often conjures images of tearing down your entire IT infrastructure and starting fresh. But let's get one thing straight: you don't need to bulldoze your existing systems to unlock the benefits of the cloud. Instead, think of it as a thoughtful remodel - you're not tearing the house down, just adding that fancy new kitchen you've always wanted.

Before starting the transformation journey, you need to assess your IT landscape to understand where the cloud fits in, where legacy systems can stay, and how they can work together.

Cataloging Your Systems: Mapping the Terrain

Imagine a family planning a cross-country vacation. They don't sell the minivan just because a new GPS app is available - they use the app to make the journey smoother. Similarly, assessing your IT systems is about finding opportunities to integrate, enhance, and extend what you already have, not necessarily replacing it all.

How to Map Your IT Terrain Thoughtfully

Inventory everything: Identify all applications and systems critical to your business. This includes core systems that run day-to-day operations and smaller tools that might not need a cloud component.

Document dependencies: Understand how systems interact so you can find potential bottlenecks and integration points. For example, if your ERP relies on an on-premises database for real-time order processing, moving that database to the cloud without addressing latency or API compatibility could disrupt operations. Identifying such dependencies ensures a smooth transition.

Assess modernization potential: For each system, evaluate whether it could benefit from cloud augmentation, hybrid deployment, or selective migration. Some systems might be perfect candidates for the cloud, while others may serve their purpose better on-premises.

Highlight areas for bolt-on innovation: Look for opportunities to add functionality, like integrating an AI-Powered Demand Forecasting & Inventory Optimization solution (e.g., Netstock in Azure) to enable Sales & Operations Planning while keeping your legacy ERP system intact.

Mapping your systems with these factors in mind ensures your roadmap is balanced, strategic, and focused on delivering value where it matters most.

Augmenting, Not Replacing: The Hybrid Advantage

The good news is that you don't have to uproot everything to enjoy the benefits of cloud technology. Think of the cloud as the Swiss Army knife of IT—it's versatile, powerful, and can be deployed exactly where you need it.

How Augmentation Can Work in Practice:

E-commerce bolt-ons: Let's say you're a manufacturer with no online sales channel. Rather than replacing your entire ERP, you could integrate a cloud-based e-commerce solution like Shopify hosted in Azure. This would allow you to sell directly to consumers without disrupting your backend.

Analytics enhancements: If your legacy systems are producing data but can't analyze it effectively, you can bolt on a cloud-based analytics platform to deliver real-time insights. For example, Amazon Redshift can connect to legacy databases via AWS Glue or Amazon Athena, providing modern BI capabilities without a full migration.

Automation add-ons: Streamline manual processes by integrating cloud-based automation tools like Power Automate to handle tasks such as invoice processing, freeing up resources and increasing efficiency.

These hybrid approaches allow you to leverage the cloud's benefits while keeping what already works intact.

Technical Debt Diagnostics: Taking the System's Pulse

Even if you're not planning a full-scale replacement of your legacy systems, understanding their health is absolutely crucial for successful cloud integration. Think of it like a doctor's check-up before starting a new fitness regime. Diagnosing technical debt helps you pinpoint where cloud solutions can add the most value and where modernization efforts are essential for a smooth transition. Ignoring technical debt can lead to unexpected costs, integration nightmares, and ultimately, a failed cloud initiative.

Here's a deeper dive into key diagnostic areas:

1. Assessing Integration Challenges: Unraveling the Web of Dependencies

- API Analysis: Examine the APIs (Application Programming Interfaces) of your legacy systems. Are they well-documented, standardized, and robust? Rigid or poorly documented APIs can create significant roadblocks when trying to connect to cloud services. Consider:
- API Availability: Does the system even have APIs? If not, integration will be significantly more complex.
- API Quality: Are the APIs well-documented and easy to use? Are they stable and reliable?
- API Compatibility: Are the APIs compatible with the technologies used by your cloud solutions?
- Dependency Mapping: Document all dependencies between your systems. Understand how they interact and which systems rely on others. This is critical for identifying potential bottlenecks and ensuring that cloud integration doesn't disrupt existing functionality. Consider tools that can automatically discover and map these dependencies.
- Data Flow Analysis: Trace the flow of data between systems. Where does the data originate? How is it transformed and used by different applications? Understanding data flows is crucial for planning data migration and integration strategies.
- Integration Points: Identify all existing integration points between your legacy systems. These points will be crucial for connecting to cloud services. Consider the technologies used for these integrations (e.g., message queues, ESBs) and their compatibility with cloud solutions.
- Integration Complexity: Assess the complexity of integrating with each system. Are there any custom integrations or unique configurations that will make cloud integration more challenging?

2. Determining Cloud Readiness: Evaluating System Architecture

- Architecture Style: Analyze the architecture of your legacy systems. Are they monolithic, or are they composed of smaller, more independent modules (e.g., microservices)? Systems with microservices or API-based architectures are generally more cloud-ready than monolithic applications.

- Technology Stack: Identify the technologies used in your legacy systems. Are these technologies supported by your chosen cloud provider? Are there any compatibility issues that need to be addressed?
- Scalability: How easily can the system scale to meet increased demand? Cloud solutions often offer auto-scaling capabilities, but your legacy systems need to be able to leverage these features.
- Maintainability: How easy is it to maintain and update the system? Systems with high levels of technical debt or outdated technologies can be difficult to integrate with the cloud.
- Security Posture: Assess the security posture of your legacy systems. Are there any known vulnerabilities? How will you ensure that security is maintained in a hybrid environment?

3. Analyzing Total Cost of Ownership (TCO): The Financial Equation

- Hardware Costs: Calculate the cost of maintaining your on-premises hardware, including servers, storage, and networking equipment.
- Software Costs: Factor in software licensing fees, maintenance costs, and support contracts.
- Operational Costs: Include costs for power, cooling, data center space, and IT staff.
- Development Costs: Estimate the cost of developing and maintaining custom integrations or middleware.
- Cloud Costs: Project the costs of using cloud services, including compute, storage, data transfer, and managed services. Don't forget to include potential cost savings from reduced hardware maintenance and operational expenses.

- Hidden Costs: Be aware of potential hidden costs, such as training, migration expenses, and the cost of managing a hybrid environment.

4. Quantifying Technical Debt: Measuring the Impact

Beyond simply identifying technical debt, try to quantify it. This can help you prioritize which systems to address first. Consider metrics like:

- Code Complexity: Measure the complexity of the codebase. Highly complex code is often more difficult to maintain and integrate.
- Code Churn: Track how often the code is changed. High code churn can indicate instability or poor design.
- Test Coverage: Assess the extent to which the code is covered by automated tests. Low test coverage increases the risk of introducing bugs during integration.
- Defect Rate: Monitor the number of defects found in the system. A high defect rate can indicate underlying problems with the code or architecture.

By thoroughly diagnosing your technical debt, you can create a realistic roadmap for cloud integration. You'll be able to identify which systems are prime candidates for cloud augmentation, which require modernization before integration, and which might be best left on-premises for the time being. This approach will save you time, money, and headaches in the long run.

Reframing Prioritization: Where Cloud Fits

Rather than treating cloud transformation as a wholesale migration project, prioritize areas where the cloud can deliver immediate and meaningful value:

- Customer-facing systems: Modernize areas where you directly interact with customers, like sales and support. For example, integrating a cloud-based chatbot with your existing help desk system can improve customer service without changing your core platform.
- Innovation accelerators: Focus on areas where cloud services can enable faster development of new capabilities, such as deploying Azure Functions to process IoT sensor data from manufacturing equipment, triggering real-time alerts and preventive maintenance actions when anomalies are detected.
- Data-driven insights: Prioritize systems that produce valuable data but lack the ability to analyze it effectively. Adding cloud-based AI or analytics capabilities can unlock new opportunities without replacing data sources.

Prioritization Framework: Focusing Your Cloud Efforts for Maximum Impact

Cloud transformation, even in a hybrid model, shouldn't be a scattershot approach. To maximize the value of your cloud investments and minimize disruption, a structured prioritization framework is essential. This framework helps you determine where cloud adoption will have the biggest impact on your business, allowing you to focus your resources strategically. We'll move beyond simply identifying where cloud could fit and delve into where it should fit first.

Here's a breakdown of key factors to consider when prioritizing your cloud initiatives:

1. Business Value: The Impact on Your Bottom Line

- Revenue Growth: Will this cloud initiative directly increase sales, expand market reach, or enable new revenue streams? For example, a cloud-based e-commerce platform integrated with your legacy ERP could unlock significant revenue potential.

- Cost Savings: Can the cloud help reduce operational costs, such as infrastructure maintenance, licensing fees, or energy consumption? A move to serverless functions for specific tasks, for instance, might offer considerable savings.
- Customer Satisfaction: Will this initiative improve the customer experience, leading to increased loyalty and retention? Think of modernizing customer support with a cloud-based chatbot or personalized recommendations powered by cloud AI.
- Operational Efficiency: Can the cloud streamline internal processes, automate manual tasks, or improve collaboration? Consider cloud-based workflow automation or document management systems.
- Risk Mitigation: Will this initiative reduce business risks, such as data loss, security breaches, or downtime? Cloud-based disaster recovery and backup solutions are prime examples.

2. Technical Feasibility: Navigating the Technical Landscape

- Integration Complexity: How easily can the proposed cloud solution integrate with your existing legacy systems? Are well-documented APIs available? Will middleware be required? Complex integrations can significantly increase project timelines and costs.
- Cloud Readiness: Is the system architecture suitable for cloud integration? Systems with microservices or API-based designs are generally easier to connect to cloud services than monolithic applications.
- Technical Debt: Does the system have significant technical debt that needs to be addressed before cloud integration can be considered? Addressing technical debt might be a prerequisite for a successful cloud initiative.

- Resource Availability: Do you have the necessary technical skills and expertise in-house to implement and manage the cloud solution? If not, will you need to invest in training, hiring, or outsourcing?

3. Risk Assessment: Identifying Potential Challenges

- Security Risks: What are the potential security implications of integrating this system with the cloud? How will you protect sensitive data in a hybrid environment?
- Data Migration Risks: If data migration is involved, what are the potential risks of data loss, corruption, or inconsistency? How will you ensure data integrity during the migration process?
- Business Disruption: What is the potential for business disruption during the implementation of the cloud solution? How can you minimize downtime and ensure business continuity?
- Vendor Lock-in: Are you overly reliant on a specific cloud vendor? How can you avoid vendor lock-in and maintain flexibility in the future?

4. Urgency: Responding to Immediate Needs

- Time Sensitivity: Is this initiative driven by a time-sensitive requirement, such as a regulatory deadline or a competitive opportunity?
- Business Criticality: Is this system essential to the core operations of your business? Prioritize cloud initiatives for mission-critical systems.
- Opportunity Window: Is there a limited window of opportunity to capitalize on this cloud initiative? For example, a new market trend or a technological advancement.
- A Practical Approach to Prioritization

- Create a Scoring Matrix: Develop a scoring matrix that assigns weights to each of the factors described above, reflecting their relative importance to your business.
- Evaluate Each Initiative: Assess each potential cloud initiative against the scoring criteria and assign a score for each factor.
- Calculate Overall Score: Calculate an overall score for each initiative by multiplying the individual scores by the corresponding weights and summing the results.
- Prioritize Based on Score: Prioritize the cloud initiatives based on their overall scores, starting with the highest-scoring projects.

By using this structured approach, you can ensure that your cloud transformation efforts are focused on the areas that will deliver the greatest value to your business, while minimizing risks and maximizing your return on investment. This approach will also allow you to communicate your cloud strategy clearly to stakeholders, justifying your decisions and building support for your initiatives.

Thinking Outside the Box: Cloud as a Bridge, Not a Replacement

The cloud isn't just about ripping and replacing your existing IT infrastructure. In many cases, it serves as a powerful bridge, connecting your valuable legacy systems to modern solutions and unlocking new capabilities without the need for disruptive and costly overhauls. This "bridge" approach allows you to innovate incrementally, minimizing disruption while maximizing the value of your existing investments. Think of it as adding a modern wing to a historic building, preserving its charm while expanding its functionality.

Here are some expanded examples of how the cloud can act as a bridge:

Automated Document Processing: From Paper to Insights

A financial services firm relies on a robust on-premises document management system, a repository of decades' worth of critical information. However, manually processing the vast volume of paper documents is slow, error-prone, and expensive. Instead of replacing the entire system, they integrate a cloud-based AI-powered Optical Character Recognition (OCR) service. This service automatically extracts and categorizes data from scanned contracts, invoices, and other documents, feeding the structured data back into the legacy system. This dramatically improves efficiency, reduces manual effort, and enables faster access to critical information, all without disrupting the core document management workflow. The cloud acts as the bridge between paper-based processes and digital insights.

AI-Driven Customer Support: Enhancing Service, Not Replacing Core Systems

A company running an on-premises CRM system wants to enhance its customer support capabilities. Instead of migrating to a new cloud-based CRM, they integrate a cloud-based AI chatbot. This chatbot can access data from the legacy CRM, enabling 24/7 automated responses to common customer inquiries. The chatbot can also escalate complex issues to human agents, who can access complete customer history from the existing CRM. This approach improves customer service, reduces response times, and frees up human agents to focus on more complex issues, all while preserving the integrity of the core customer records within the on-premises system. The cloud AI acts as the bridge to 24/7 support and improved customer satisfaction.

Real-time Analytics for Legacy Data: Unlocking Hidden Potential

A manufacturing company has a legacy system that collects vast amounts of production data. However, this data is only used for basic reporting. By integrating a cloud-based analytics platform, they can unlock the hidden potential of this data. The cloud platform connects to the legacy system, extracts the data, and applies advanced analytics techniques to identify trends, patterns, and anomalies. This enables real-time insights into production performance, allowing them to optimize processes, predict equipment failures, and improve product quality, all without replacing the legacy data source. The cloud analytics platform acts as the bridge to valuable business intelligence.

Extending Functionality with Serverless Computing: Agile Innovation

A retail company uses a legacy order management system. They want to add a new feature: real-time inventory updates triggered by sales transactions. Instead of modifying the complex legacy system, they use serverless computing functions in the cloud. When a sale occurs, the legacy system triggers a cloud function, which updates the inventory in a separate cloud-based database. This approach allows them to add new functionality quickly and easily without touching the core legacy system. The serverless functions act as the bridge to agile innovation.

API-led Connectivity: The Foundation for Hybrid Solutions

Many legacy systems lack modern APIs, making integration with cloud services challenging. Creating a layer of APIs around these systems can be a crucial first step. This can be achieved through API gateways or by developing custom adapters. Once APIs are in place, it becomes much easier to connect the legacy system to cloud services, enabling a wide range of hybrid solutions. These APIs act as the essential bridge for future cloud integrations.

Benefits of the "Bridge" Approach

Reduced Disruption: Incremental changes minimize disruption to existing business processes.

- Lower Cost: Avoids the significant expense of full-scale system replacements.
- Faster Time to Value: New capabilities can be deployed more quickly.
- Reduced Risk: Smaller, more focused projects reduce the risk of failure.

Preservation of Existing Investments: Leverages the value of existing legacy systems.

By embracing the "cloud as a bridge" mentality, organizations can embark on their cloud journey at their own pace, focusing on the areas that deliver the most immediate value. This pragmatic approach enables innovation without disruption, maximizing the return on existing IT investments while paving the way for future modernization efforts.

Use Case Example – AI for legacy data:

A healthcare provider faces the challenge of leveraging its legacy patient record system, an old but reliable platform that contains decades of valuable data. Instead of replacing the system, which would be costly and disruptive, they integrate a cloud-based AI analytics tool to unlock the potential of their historical data.

The AI platform connects seamlessly with the legacy system, processing millions of records to identify trends and patterns. For instance, it uses historical data to predict patient readmission risks, highlight anomalies in treatment outcomes, and recommend preventative measures for chronic disease management. The AI tool's ability to generate predictive insights empowers clinicians to make data-driven decisions while continuing to rely on their familiar record-keeping system.

This solution bridges the gap between traditional IT infrastructure and cutting-edge innovation, allowing the healthcare provider to modernize its capabilities without overhauling its core systems. The result is improved patient care, enhanced operational efficiency, and cost savings—achieved through a strategic combination of legacy stability and AI-driven foresight.

Conclusion: It's Not All or Nothing

Assessing your legacy landscape isn't just about deciding what to migrate—it's about discovering where the cloud can add the most value. By embracing a hybrid mindset, you can combine the best of both worlds: the stability of your legacy systems and the flexibility of modern cloud solutions.

Cloud transformation is like renovating your house room by room. Start with the areas that deliver the most impact—whether it's bolting on an e-commerce platform, adding analytics capabilities, or automating routine tasks. This pragmatic approach ensures that your IT transformation drives real value while minimizing disruption.

So, let go of the "rip and replace" mindset. With a thoughtful assessment and a focus on hybrid solutions, you can unlock the cloud's full potential while keeping what works—and leaving the wrecking ball in storage.

Chapter 5: Choosing the Right Cloud Strategy

Migrating to the cloud isn't a simple "lift and shift" operation. It's a strategic undertaking, akin to commissioning a bespoke suit. It needs to fit your business perfectly, reflect your unique style, and be tailored for optimal comfort and performance. This chapter serves as your comprehensive guide, navigating the intricate landscape of cloud models, migration approaches, and the strategic considerations surrounding augmenting or replacing your legacy systems.

By the end of this chapter, you'll be equipped with the knowledge and insights to confidently formulate your cloud game plan.

Cloud Models 101: Selecting the Right Tool for the Job

The world of cloud models isn't just about choosing between a sedan and an SUV; it's like stepping into a vast showroom with a diverse range of vehicles, each possessing its own distinct features and capabilities. Selecting the right model requires a deep understanding of what's under the hood – the specific needs of your business.

Public Cloud:

Imagine carpooling on a luxurious, well-maintained bus. Providers like AWS, Azure, and Google Cloud manage the entire operation, offering a diverse menu of services accessible to anyone with a ticket. This model provides flexibility, scalability, and often, a lower initial cost. However, if your cargo is exceptionally sensitive – think personal health information or confidential financial records – you might have concerns about sharing the ride. You'll need to carefully evaluate the security and compliance offerings of the public cloud provider.

Private Cloud:

This is your personal limousine. You have complete control over the vehicle, the route, and the passenger list. A private cloud is ideal for organizations with stringent security and compliance requirements. However, this level of control comes with a higher price tag, and you're responsible for the ongoing maintenance and upkeep.

Example Use Case of Private Cloud

Scenario: "Global Bank," a large financial institution, needs to modernize its core banking system to meet escalating customer demands and adhere to strict regulatory requirements.

Challenge: Global Bank's legacy system processes highly sensitive financial data, demanding the utmost security and control. Migrating to a public cloud raises concerns about data privacy, regulatory compliance, and potential vulnerabilities.

Solution: Global Bank chooses a private cloud solution. They establish a dedicated cloud environment within their own data centers, giving them complete command over their infrastructure, data, and security protocols.

Benefits:

- Enhanced Security: Global Bank can implement stringent security protocols and access controls tailored to their specific regulatory obligations. This minimizes the risk of data breaches and ensures compliance with industry standards.
- Data Privacy: Sensitive customer data remains within Global Bank's private cloud, reducing exposure to external threats and guaranteeing adherence to data privacy regulations.
- Customization: Global Bank can tailor the private cloud environment to perfectly match the performance and configuration needs of their core banking application.
- Control: They have complete control over their infrastructure, allowing them to optimize resource allocation, implement

robust disaster recovery measures, and maintain high availability for critical banking services.

In essence, Global Bank leverages a private cloud to modernize its core banking system while preserving the highest levels of security, privacy, and control over its sensitive financial data. This strategy enables them to innovate and meet customer demands without compromising on compliance or security.

Hybrid Cloud:

Meet the hybrid SUV of cloud computing. This model combines the power and security of a private cloud for your most sensitive data with the agility and cost-effectiveness of a public cloud for other workloads. Need the best of both worlds? The hybrid cloud might be your ideal solution.

Example Use Case of Hybrid Cloud

Scenario: "Trendy Threads," a rapidly expanding e-commerce company, needs a scalable and cost-effective solution to manage fluctuating website traffic and data storage requirements.

Challenge: Trendy Threads experiences substantial spikes in website traffic during peak seasons and promotional events. Their existing on-premises infrastructure struggles to handle these surges, leading to slowdowns and potential customer dissatisfaction. They also require a secure and compliant solution for storing sensitive customer data.

Solution: Trendy Threads adopts a hybrid cloud strategy. They maintain their on-premises data center for sensitive customer data and core applications, ensuring compliance and control. They leverage a public cloud provider for:

Scalable Web Hosting: During peak traffic periods, they "burst" their web traffic to the public cloud, automatically scaling their web servers to accommodate the increased demand. This guarantees a seamless customer experience even during peak shopping seasons.

Data Analytics and Machine Learning: They utilize public cloud services for data warehousing and analytics, processing large datasets to gain insights into customer behavior, optimize marketing campaigns, and personalize product recommendations.

Disaster Recovery: They replicate critical data and applications to the public cloud as a backup, ensuring business continuity in case of an on-premises outage.

Benefits:

- Cost Optimization: Trendy Threads avoids over-provisioning on-premises infrastructure by using the public cloud for scalable resources. They pay only for what they use during peak periods.
- Scalability and Performance: The public cloud provides on-demand scalability to handle traffic spikes, ensuring optimal website performance and customer satisfaction.
- Security and Compliance: Sensitive data remains secure in their private cloud, while less critical data and applications can leverage the public cloud's security features.
- Innovation: They can access a wide range of public cloud services, such as machine learning and analytics, to drive innovation and gain a competitive edge.

In essence, Trendy Threads strategically combines the security and control of a private cloud with the scalability and cost-effectiveness of a public cloud. This hybrid approach allows them to adapt to changing demands, optimize costs, and leverage cloud innovation while maintaining data security and compliance.

Multi-Cloud:

Ever been to an all-you-can-eat buffet? Multi-cloud is the IT equivalent. You select the best offerings from each vendor's menu, avoiding vendor lock-in and optimizing costs. However, managing this diverse array of services can become complex if you're not well-organized.

Example Use Case of Multi-Cloud

Scenario: "Media Giant," a global media and entertainment company, needs a robust and flexible cloud strategy to support its diverse services, including video streaming, online gaming, and content production.

Challenge: Media Giant requires a cloud solution that can handle massive data storage and processing needs, ensure low latency for its global user base, and provide access to cutting-edge technologies in different domains. Relying on a single cloud provider poses risks of vendor lock-in, limited geographic reach, and potential service disruptions.

Solution: Media Giant adopts a multi-cloud strategy, leveraging the strengths of different cloud providers:

AWS for Video Streaming: They utilize AWS's extensive content delivery network (CDN) and media services to deliver high-quality video streaming to its global audience with low latency and high availability.

Google Cloud for AI and Machine Learning: They leverage Google Cloud's expertise in AI and machine learning for content analysis, personalized recommendations, and fraud detection.

Azure for Gaming Infrastructure: They utilize Azure's gaming services and global infrastructure to host and scale their online gaming platform, providing a seamless gaming experience for users worldwide.

Benefits:

- Avoids Vendor Lock-in: Media Giant reduces dependency on a single provider, gaining flexibility and negotiating power.
- Optimizes Costs: They can choose the most cost-effective services from each provider, leveraging competitive pricing and avoiding overspending.

- Access to Best-of-Breed Solutions: They leverage the specific strengths of each provider, gaining access to cutting-edge technologies in different domains.
- Improved Resilience and Availability: Distributing workloads across multiple clouds reduces the risk of service disruptions and ensures high availability for critical services.
- Enhanced Global Reach: They can leverage the global infrastructure of different providers to optimize content delivery and service availability for users worldwide.

In essence, Media Giant strategically utilizes multiple cloud providers to create a robust, flexible, and cost-effective cloud ecosystem. This multi-cloud approach allows them to optimize performance, leverage specialized services, and mitigate risks while supporting their diverse media and entertainment offerings.

My Opinion

Most businesses will thrive in the public cloud. Don't get bogged down worrying about data security for your everyday needs - the major public cloud providers have security covered. Focus on your core business, not on building and managing complex IT infrastructure.

The Takeaway

There's no "wrong" answer here, just a matter of what fits your budget, your compliance needs, and the way your business operates. Don't fall into the trap of choosing what's popular - choose what works for you.

Migration Approaches: Finding the "Just Right" Solution

Once you've chosen your ideal cloud model, it's time to chart your course to the cloud. This stage of your journey presents three distinct paths: lift-and-shift, re-platforming, and rebuilding. Like Goldilocks tasting porridge, you need to find the approach that's "just right" for your applications and business objectives.

Lift-and-Shift:

This is the "pack up the moving truck and go" approach. You migrate your applications to the cloud without modifying their underlying architecture or code. It's the fastest and often the most cost-effective option upfront, but you're essentially bringing your existing baggage along. While it might fit into your new cloud environment, it may not be optimized for cloud-native capabilities.

Example Use Case for a Lift-and-Shift Cloud Migration

Scenario: "Bookworm Books," a mid-sized retail company, wants to quickly move its e-commerce website to the cloud to improve scalability and reduce infrastructure costs.

Challenge: Bookworm Books' online store experiences occasional performance issues during peak shopping seasons and promotional events. Their on-premises servers are nearing capacity, and upgrading their hardware would be expensive and time-consuming. They need a quick and cost-effective solution to improve website performance and scalability.

Solution: Bookworm Books chooses a lift-and-shift migration strategy. They replicate their existing website infrastructure, including the operating system, web server, and database, to a public cloud provider (e.g., AWS, Azure, or Google Cloud). This involves minimal code changes and can be accomplished relatively quickly.

Benefits:

- Speed: The migration is completed quickly, minimizing disruption to the online business and allowing them to realize benefits sooner.
- Cost Reduction: They eliminate the need to invest in new hardware and reduce ongoing maintenance costs by leveraging the cloud provider's infrastructure.
- Improved Scalability: The public cloud provides on-demand scalability, allowing them to handle traffic spikes during peak seasons without performance issues.

- Reduced IT Burden: They offload infrastructure management to the cloud provider, freeing up their IT team to focus on other priorities.

In essence, Bookworm Books "lifts" their existing e-commerce website from their on-premises servers and "shifts" it to the cloud with minimal changes. This approach allows them to quickly gain the benefits of cloud computing without the time and cost associated with more complex migration strategies.

My Opinion

Lift-and-shift is often a suitable starting point for cloud migration, especially for applications that are not overly complex or tightly coupled to on-premises infrastructure. It provides a quick win and allows businesses to experience the benefits of the cloud without significant upfront investment. However, it's essential to recognize that lift-and-shift may not fully optimize cloud capabilities and could result in higher long-term costs if not followed by further optimization efforts.

Re-Platforming:

Think of this as upgrading your furniture before moving into a new home. You make some modifications to your applications – perhaps replacing an outdated database or operating system – to make them more compatible with the cloud environment. It requires more effort than lift-and-shift, but you're setting yourself up for a smoother and more optimized cloud experience.

Example Use Case for Re-Platforming Cloud Migration

Scenario: "HealthyLife Clinic," a healthcare organization, wants to modernize its patient management system to improve efficiency, scalability, and accessibility for its staff.

Challenge: HealthyLife Clinic relies on an aging on-premises application for managing patient records, scheduling appointments, and billing. The system is becoming increasingly difficult to maintain, lacks mobile access for doctors and nurses, and struggles to handle the growing number of patients.

Solution: HealthyLife Clinic opts for a re-platforming migration strategy. They move the application to a public cloud infrastructure (e.g., AWS, Azure, or Google Cloud) and make some modifications to optimize it for the cloud environment. This might involve:

Switching to a Managed Database Service: They migrate their on-premises database to a cloud-managed database service (e.g., Amazon RDS, Azure SQL Database, or Google Cloud SQL) to improve scalability, reliability, and reduce administrative overhead.

Updating the Application's User Interface: They modernize the application's user interface to make it more user-friendly and accessible on mobile devices, enabling doctors and nurses to access patient information and update records from anywhere.

Implementing Automated Scaling: They configure the application to automatically scale resources based on demand, ensuring optimal performance during peak hours and reducing costs during periods of low activity.

Benefits:

- Improved Performance and Scalability: The cloud-optimized application can handle increased workloads and provide a better user experience for staff.
- Enhanced Accessibility: Mobile access empowers healthcare professionals to access patient information and update records from anywhere, improving efficiency and patient care.
- Reduced IT Burden: Using managed services reduces the administrative burden on the IT team, allowing them to focus on other priorities.
- Cost Optimization: They can leverage cloud pricing models to optimize costs and pay only for the resources they consume.

In essence, HealthyLife Clinic "lifts" its patient management system, "tunes" it to make it cloud-friendly, and then "shifts" it to the cloud. This re-platforming approach allows them to modernize their application and take advantage of cloud benefits without a complete rewrite, striking a balance between cost, speed, and optimization.

My Opinion:

Re-platforming is a prudent choice when you want to modernize an application without the complexity and cost of a full rebuild. It allows you to leverage cloud benefits while retaining the core functionality of your existing system. HealthyLife Clinic's example illustrates how re-platforming can improve efficiency, scalability, and accessibility for critical applications. However, it's important to carefully assess the application's architecture and dependencies to ensure a smooth transition and avoid unexpected challenges during the re-platforming process.

Rebuilding:

If the first two approaches are akin to patching and repairing, rebuilding is a complete renovation. You dismantle your legacy applications and rebuild them from the ground up using cloud-native technologies and architectures. This approach offers unmatched performance, scalability, and flexibility, but it's also the most expensive and time-consuming option.

Example Use Case for Rebuilding in Cloud Migration

Scenario: "Game On!" is a popular online gaming company that wants to revamp its aging gaming platform to incorporate modern features, improve scalability, and enhance the user experience.

Challenge: Game On!'s current platform is built on outdated technology and struggles to handle the increasing number of players and the demand for new features like social interaction, in-game purchases, and cross-platform compatibility. Simply migrating the existing platform to the cloud wouldn't address these challenges effectively.

Solution: Game On! decides to rebuild its gaming platform from scratch using cloud-native technologies and a microservices architecture. This involves:

- Choosing a Cloud-Native Technology Stack: They select modern programming languages, frameworks, and databases that are optimized for cloud environments and scalability.
- Adopting a Microservices Architecture: They break down the platform into smaller, independent services that can be developed, deployed, and scaled independently. This allows for greater flexibility, faster development cycles, and improved fault tolerance.
- Leveraging Cloud-Native Services: They utilize managed services offered by their chosen cloud provider (e.g., AWS, Azure, or Google Cloud) for databases, messaging, storage, and other functionalities, reducing operational overhead and improving efficiency.
- Designing for Scalability and Resilience: They architect the new platform to handle massive player concurrency, ensure high availability, and automatically scale resources based on demand.

Benefits:

- Enhanced User Experience: The new platform offers a modern, engaging experience with improved features, performance, and stability.
- Increased Agility and Innovation: The microservices architecture allows for faster development and deployment of new features and updates, enabling Game On! to stay ahead of the competition.
- Improved Scalability and Reliability: The cloud-native design and managed services ensure the platform can handle massive player growth and maintain high availability.

- Cost Optimization: They can leverage cloud pricing models and resource optimization techniques to manage costs effectively.

In essence, Game On! "tears down" its old platform and "rebuilds" a new one from the ground up, specifically designed for the cloud environment and optimized for modern gaming demands. This rebuilding approach allows them to overcome the limitations of their legacy system, embrace innovation, and deliver a superior gaming experience to their users.

Joel's Take: Matching Strategy to Business Value

Rebuilding is the most complex and resource-intensive migration strategy, but it offers the greatest potential for long-term benefits. Game On!'s example demonstrates how rebuilding can unlock innovation, improve scalability, and enhance the user experience. However, it requires careful planning, a skilled development team, and a strong understanding of cloud-native technologies and architectures. It's crucial to have a clear vision, a well-defined roadmap, and a commitment to ongoing innovation to ensure a successful rebuilding effort. The Rebuilding approach is best suited for proprietary intellectual property, like in the use case of Game On!. I would avoid a Rebuild strategy for any legacy solution that can be replaced by a modern Commercially Off The Shelf (COTS) solution like CRM or MRP.

The Takeaway:

The optimal migration path depends on several factors: the complexity of your applications, your budget and timeline constraints, your risk tolerance, and your long-term business goals. Sometimes, a simple lift-and-shift is sufficient; other times, a complete rebuild is necessary to fully realize the benefits of the cloud. Carefully evaluate your resources, goals, and the specific needs of your applications to make an informed decision.

Choosing the right cloud strategy isn't merely a technical decision—it's a strategic pivot point, setting the trajectory for your organization's future growth, agility, and competitive edge. As we've explored, the cloud isn't a one-size-fits-all scenario; it's a tailored journey that demands careful consideration of your unique business needs, technical realities, and long-term objectives.

Whether your path involves the simplicity of lift-and-shift, the moderate enhancements of re-platforming, or the comprehensive innovation of a full rebuild, each approach offers distinct benefits and trade-offs. Ultimately, your goal is clarity—understanding precisely why you're migrating to the cloud and aligning your approach accordingly.

But knowing the right cloud strategy is just the beginning. How do you move from strategic intent to practical execution? How can you ensure your journey stays aligned with your business goals, avoids pitfalls, and delivers the anticipated outcomes?

The answers lie ahead. In the next chapter, we'll explore a structured framework—the Cloud Transformation Roadmap - that guides your organization step-by-step through planning, execution, and optimization of your cloud journey. It's your actionable playbook, ensuring your strategic decisions translate into real-world success.

Let's embark on this next stage of the journey together.

Chapter 6: The Cloud Transformation Roadmap

Migrating to the cloud isn't a quick weekend project - it's a thoughtful journey that unfolds in phases. Each step builds on the last, ensuring you're set up for long-term success. In this chapter, we're diving into the four essential phases of cloud transformation: Discovery and Planning, Proof of Concept, Implementation and Optimization, and Continuous Improvement. Think of this roadmap as your trusty guide to navigating the sometimes tricky, but always rewarding, path to the cloud.

Phase 1: Discovery and Planning – Laying the Foundation

Every great journey starts with a clear map. Before making your first move toward the cloud, you need to know your starting point, your destination, and the best way to get there.

Assess Your Current State

Inventory your IT landscape like a spring-cleaning session. What applications and data are you working with? Are there bottlenecks or technical debt dragging you down? Map out your pain points and identify what's ripe for improvement.

Define Business Objectives

Why are you moving to the cloud? Be specific: are you looking for scalability, cost reduction, faster innovation, or all of the above? Define what success means to your business, and don't be afraid to dream big.

Choose a Cloud Strategy

This is where you define the structure of your cloud approach. Will it be public, private, hybrid, or multi-cloud? Do you go for a lift-and-shift approach, or rebuild your applications from scratch for the cloud? Each choice has trade-offs, so align it with your business goals.

Build a Migration Plan
A solid plan is your safety net. Outline timelines, allocate resources, and set budgets. Prioritize which apps to migrate first, identify risks, and establish success metrics.

Secure Executive Buy-In
Leadership support is critical. Build a compelling case by showcasing the ROI and strategic benefits of cloud transformation. Get stakeholders excited and onboard.

Key Takeaway: Start with clarity. With a well-defined strategy and a solid plan, you'll set yourself up for a smooth transition to the cloud.

Use Case Example for Phase 1: Discovery and Planning

Fictional Organization: "Precision Parts Co." - a mid-sized manufacturer of specialized components for the automotive and aerospace industries.

Profile: Precision Parts Co. has been operating for over 50 years and relies heavily on legacy systems for design, production, inventory management, and customer relations. These systems are becoming increasingly difficult and expensive to maintain, hindering their ability to innovate and compete in a rapidly evolving market.

Phase 1: Discovery and Planning - Precision Parts Co.'s Journey

The leadership team at Precision Parts Co. recognized the need for change and embarked on a cloud transformation journey. Here's how they approached Phase 1:

Assess Your Current State: They conducted a thorough assessment of their IT landscape, documenting all applications, systems, and data flows. They identified several pain points, including outdated hardware, slow application performance, limited scalability, and a growing backlog of IT maintenance tasks.

Define Business Objectives: Precision Parts Co. defined clear objectives for their cloud transformation:

Improve operational efficiency: Streamline processes, reduce manual effort, and improve productivity.

Enhance scalability: Accommodate growing data volumes and support future expansion plans.

Reduce IT costs: Lower infrastructure and maintenance costs while optimizing resource utilization.

Foster innovation: Enable faster development and deployment of new applications and services to support business growth.

Choose a Cloud Strategy: After careful consideration, they opted for a hybrid cloud strategy. They decided to retain their on-premises infrastructure for sensitive design data and core manufacturing applications, while leveraging a public cloud provider for:

Scalable data storage and analytics: Storing and analyzing large datasets generated by their production processes to optimize efficiency and quality control.

Customer relationship management (CRM): Implementing a cloud-based CRM system to improve customer interactions and support sales growth.

Disaster recovery: Replicating critical data and applications to the public cloud for business continuity and disaster recovery purposes.

Build a Migration Plan: They developed a detailed migration plan, outlining timelines, resource allocation, and budget considerations. They prioritized migrating their CRM system and data analytics platform to the cloud first, followed by a phased migration of other applications.

Secure Executive Buy-In: The IT team presented a compelling business case to the executive leadership, highlighting the potential ROI, cost savings, and strategic benefits of cloud transformation. They secured full support from the leadership team to proceed with the migration.

Key Takeaway: Precision Parts Co. laid a strong foundation for their cloud transformation by thoroughly assessing their current state, defining clear objectives, choosing a suitable cloud strategy, developing a comprehensive migration plan, and securing executive buy-in. This methodical approach set them up for success in the subsequent phases of their cloud journey.

Phase 2: Proof of Concept – Testing the Waters

Think of the proof of concept (POC) as your dress rehearsal. This is where you test assumptions, work out the kinks, and get a feel for how the cloud will operate in your specific environment.

Select a Pilot Project
Choose an application or workload that's important but not mission-critical. Think of it as the sandbox where you can experiment without high stakes.

Set Measurable Goals
Create SMART goals (Specific, Measurable, Achievable, Relevant, Time-bound) for your POC. For example, you might aim to improve application response time by 20% or cut infrastructure costs by 15%.

Implement and Monitor
Execute the migration of your pilot project and keep a close watch on its performance. Use monitoring tools to gather data on key metrics.

Evaluate and Refine

Analyze the results. What worked? What didn't? Use the lessons learned to refine your approach and inform the larger rollout.

Detailed Proof of Concept Example 1: Re-Platform for a Retail Giant

Imagine a large retail company, RetailX, looking to modernize its online operations with a cloud migration. To dip their toes in the water, they select their loyalty rewards program as the POC. Why? It's important enough to demonstrate value but won't halt operations if the migration stumbles.

Here's how RetailX's POC unfolds:

- Defining Goals - RetailX sets clear objectives for the POC:
  - Reduce database query times by 30% for loyalty rewards redemptions.
  - Enhance scalability to handle peak holiday traffic.
  - Cut infrastructure costs by 20% compared to their on-premises solution.
- Pilot Execution
  - RetailX migrates the loyalty program's database to a managed cloud database service while shifting the application layer to cloud-based virtual machines.
- Monitoring Metrics
  - They closely monitor performance metrics, such as database response times, user login speeds, and overall system availability. They also track cost savings and customer satisfaction scores via surveys.
- Iterating and Improving-During the POC, they identify areas for improvement:
  - The cloud database initially struggles with a specific query structure, leading to a tweak in how the application retrieves data.

- o Scalability tests reveal the need for additional load-balancing configurations.
- o After adjustments, RetailX exceeds their goals, cutting query times by 35%, achieving scalability for peak periods, and reducing costs by 25%.

- Key Insights Gained
  - o RetailX learns that re-platforming databases is more effective than lift-and-shift in their environment. They also identify training needs for their IT team to manage cloud-based workflows effectively.

Detailed Proof of Concept Example 2: Hybrid Cloud for a CPG Company

Let's imagine a Consumer-Packaged Goods (CPG) company, FreshPoint Foods, that's grappling with the age-old nemesis of data silos. Their on-premises systems are scattered across different functions - sales, supply chain, production, and customer service - leaving their data fragmented and underutilized. Leadership sees an opportunity to use the hybrid cloud for data consolidation, analytics, and AI-driven insights.

Objective: FreshPoint wants to centralize their data and leverage advanced analytics to improve operational efficiency. To start small, they decide to conduct a POC focused on demand forecasting within their supply chain.

Why Demand Forecasting?

Demand forecasting is critical in the CPG industry, where overstocking leads to waste, and understocking means missed sales. FreshPoint believes that by centralizing their supply chain data and applying predictive analytics, they can:

- o Reduce forecast errors by 20%.
- o Improve inventory turnover by 15%.
- o Minimize out-of-stock situations for high-demand products.

- FreshPoint's Proof of Concept Plan

Step 1: Define POC Goals

FreshPoint identifies specific, measurable objectives for the demand forecasting POC:

- Consolidate sales, production, and distribution data from three key product categories (e.g., beverages, snacks, and frozen meals).
- Use AI models to predict demand for the next six months.
- Improve forecast accuracy compared to their current manual Excel-based process.

Step 2: Select the Hybrid Cloud Environment

- Given their on-prem systems and the need for secure, scalable analytics, FreshPoint opts for a hybrid cloud strategy.
- On-premise: Their existing ERP and POS systems remain on-prem, feeding data into the cloud.
- Cloud: A public cloud platform is used for data storage, analytics, and AI modeling.

Step 3: Build the Data Pipeline

- FreshPoint sets up a data pipeline to consolidate disparate sources:
- Sales data from retail partners via their on-prem POS system.
- Production data from their ERP system tracking manufacturing schedules.
- Distribution data from their logistics software.
- The data is ingested into a cloud-based data lake, with connectors ensuring real-time updates.

Step 4: Develop Analytics and AI Models

- FreshPoint leverages a cloud-based AI platform to run predictive models.
- Historical sales patterns, seasonality, and promotional events are factored in to forecast demand for each product category.

Step 5: Execute and Monitor the POC

- The POC focuses on three key regions where FreshPoint has historically faced stockouts during peak periods.
- Monitoring metrics:
- Forecast accuracy compared to actual sales.
- Reduction in out-of-stock situations.
- Improvements in inventory turnover.

Step 6: Evaluate Results and Adjust

After three months of running the POC, FreshPoint evaluates the outcomes:

- Demand forecasting accuracy improves by 22%, exceeding the 20% goal.
- The beverage category shows the highest improvement, reducing stockouts by 30%.
- Inventory carrying costs decrease due to better alignment with actual demand.

Key Insights from the POC

FreshPoint learns several valuable lessons:

- **Value of Centralized Data:** Consolidating data from disparate systems into a hybrid cloud environment provides visibility they never had before.
- **Scalability of the Cloud:** The ability to process massive amounts of data in the cloud proves critical for running AI models efficiently.
- **Change Management Needs:** The supply chain team requires training to interpret and act on AI-driven forecasts.

Next Steps for FreshPoint

Encouraged by the success of their demand forecasting POC, FreshPoint plans to:

- Expand the hybrid cloud platform to include customer insights and marketing data.
- Develop AI-driven insights for other functional areas, such as pricing optimization and supplier management.
- Gradually migrate more on-prem systems to the cloud to increase agility while retaining control over mission-critical operations on-premises.

Key Takeaway for Hybrid Cloud POCs

This POC demonstrates how hybrid cloud models can address real-world challenges like data silos while offering the flexibility of cloud analytics and AI. Starting with a targeted functional area, like demand forecasting, allows organizations to validate the value of cloud capabilities while minimizing risk.

A hybrid cloud approach lets you keep one foot in the familiar while stepping into the future - ideal for companies with on-prem systems looking to modernize gradually. FreshPoint's POC didn't just consolidate data; it consolidated confidence in their cloud transformation journey.

Detailed Proof of Concept Example 3: Multi-Cloud for a Media and Entertainment Company

Let's reintroduce Visionary Studios, a global media and entertainment powerhouse specializing in video streaming, digital content production, and live event broadcasting. With a massive global audience, they've embraced a multi-cloud strategy to improve performance, reduce latency, and ensure a seamless viewing experience for users worldwide.

Objective: Visionary Studios wants to enhance their content delivery network (CDN) for live streaming. By leveraging the strengths of two public cloud providers - Amazon Web Services (AWS) and Google Cloud Platform (GCP) - they aim to:

- Minimize latency during peak traffic events.

- Ensure high availability by routing traffic between providers.
- Optimize costs with real-time pricing strategies.

Why Focus on CDN for the POC?

In the media and entertainment world, buffering or latency is a dealbreaker. For Visionary Studios, delivering flawless live-streaming experiences is essential to maintaining user loyalty and brand reputation. By testing a multi-cloud CDN, they aim to:

- Improve global content delivery: Use cloud regions and edge locations from AWS and GCP to bring content closer to users.
- Enhance resiliency: Route traffic dynamically to avoid outages and ensure uptime during live events.
- Control costs: Leverage competitive pricing between providers to optimize spend.

Visionary Studios' Proof of Concept Plan

Step 1: Define POC Goals

- Visionary Studios sets measurable goals for the POC:
- Reduce latency by 30% compared to their current single-cloud setup.
- Achieve 99.99% uptime across all regions.
- Reduce CDN costs by 15% using real-time pricing optimization.

Step 2: Select the Multi-Cloud Providers

The company chooses AWS and GCP as their multi-cloud providers, leveraging their respective strengths:

- AWS CloudFront CDN: Known for its global presence and low-latency edge locations, especially in North America and Europe.
- Google Cloud CDN: Offers competitive pricing and strong infrastructure in Asia-Pacific, with machine learning integrations for dynamic optimization.

Step 3: Architect the Multi-Cloud CDN Solution

Visionary Studios designs a solution to maximize performance and cost efficiency:

- o Dynamic Load Balancer: Traffic is routed using an intelligent multi-cloud load balancer. This tool evaluates factors like proximity, latency, and cost in real-time.
- o Content Replication: Video content is pre-replicated to both AWS and GCP regions to ensure availability.
- o Monitoring Tools: Real-time analytics tools track latency, uptime, and cost metrics across providers.

Step 4: Execute the POC During a Live Event

Visionary Studios selects a globally broadcast esports tournament as the testing ground. The event involves millions of concurrent viewers from North America, Europe, and Asia-Pacific.

During the Event:

Traffic Routing:

- o AWS CloudFront handles North American and European traffic, leveraging its extensive edge network.
- o Google Cloud CDN manages Asia-Pacific traffic, where its regional infrastructure offers lower latency.

Dynamic Pricing Optimization:

- o When AWS costs spike due to peak traffic, traffic is dynamically rerouted to GCP nodes in regions where latency differences are negligible.

Real-Time Monitoring:

- o Tools like AWS CloudWatch and Google Cloud Operations Suite monitor performance metrics and ensure smooth operations.

Step 5: Evaluate Results

After the event, Visionary Studios reviews the outcomes:

- Latency Reduction: Latency drops by an average of 35%, surpassing the 30% target. The Asia-Pacific region sees the largest improvement due to GCP's infrastructure.
- High Availability: Uptime achieves an impressive 99.999%, with zero outages despite record-breaking viewer numbers.
- Cost Savings: CDN costs are reduced by 17%, exceeding the 15% goal, thanks to real-time routing based on pricing differences.

Key Insights from the POC

The POC reveals valuable lessons for Visionary Studios:

- Provider Specialization Matters: AWS delivers exceptional performance in regions where it has a strong presence, while GCP shines in Asia-Pacific. Leveraging their strengths maximizes user experience.
- Dynamic Pricing Optimization Works: Using real-time data to route traffic based on cost saves significant money without compromising performance.
- Multi-Cloud Improves Resiliency: The ability to switch providers seamlessly during traffic surges ensures uninterrupted service.

Next Steps for Visionary Studios

Buoyed by the POC's success, Visionary Studios plans to:

- Expand the multi-cloud CDN to include video-on-demand (VOD) services.
- Integrate AI-driven traffic forecasting to preemptively route traffic during spikes.
- Explore additional cloud providers, such as Microsoft Azure, to further diversify their infrastructure.

Key Takeaway for Multi-Cloud POCs

This POC underscores the strategic value of a multi-cloud approach for global businesses. Visionary Studios demonstrated how AWS and

GCP's strengths could be combined to deliver a superior user experience, ensure uptime, and optimize costs. Multi-cloud isn't just about avoiding vendor lock-in - it's about unlocking your business's potential. Visionary Studios proved that with the right strategy, you can deliver world-class performance while keeping your operations resilient and cost-effective.

Final Thought on POCs: A well-executed POC doesn't just validate your strategy - it also provides a roadmap for scaling up and avoiding pitfalls in the next phase.

Phase 3: Implementation and Optimization – Scaling Up

With the POC in the rearview mirror, it's time to scale your efforts. This phase is all about executing your plan at a larger scale while fine-tuning along the way.

Migrate Applications and Data

- o Stick to a phased approach. Migrate workloads in batches, starting with less complex systems. This reduces risk and gives you room to adapt as needed.

Optimize for Performance and Cost

- o Cloud environments need constant tuning. Use performance monitoring tools to track metrics, and leverage cost-management tools to avoid surprises on your bill.

Training and Enablement

- o Equip your teams with the skills they need to thrive in a cloud-first world. Offer certifications, hands-on training, and clear documentation.

Change Management and Communication

- o Keep the organization informed. Regular updates, clear goals, and a steady drumbeat of communication help align teams and ease resistance to change.

Key Takeaway: This is the heavy-lifting phase. Success depends on careful coordination, constant optimization, and a proactive approach to managing both technology and people.

Phase 4: Continuous Improvement – Staying Ahead of the Curve

The finish line? It doesn't exist. Cloud transformation is a marathon with no end, but the beauty lies in the journey itself.

Monitor and Optimize

- Regularly assess your cloud environment for performance, costs, and security. Make tweaks to keep everything running smoothly.

Innovate and Adopt New Tools

- The cloud is constantly evolving. Stay curious and experiment with new services, whether it's AI-driven analytics, serverless computing, or advanced automation.

Prioritize Security and Compliance

- Cybersecurity is a moving target. Stay vigilant with strong protocols, regular audits, and compliance updates.

Foster a Learning Culture

- Encourage your team to embrace continuous learning. Certifications, webinars, and hands-on projects keep skills sharp and teams ready for the next challenge.

Key Takeaway: The cloud journey is ongoing. Stay nimble, keep innovating, and let continuous improvement drive your success.

Conclusion: A Roadmap to Cloud Success

Cloud transformation isn't just about technology—it's about evolving your business to meet the demands of tomorrow. By following this roadmap, testing your assumptions, and embracing continuous improvement, you'll unlock the full potential of the cloud. The cloud isn't the destination; it's the vehicle to get you where you want to go.

However, before you embark on this journey, it's essential to address the critical aspect of data management in the cloud. Data is the lifeblood of any organization, and migrating to the cloud presents both opportunities and challenges for data handling. In the next chapter, "**Addressing Data Challenges**," we'll delve into the intricacies of data migration, storage, security, and governance in the cloud, equipping you with the knowledge and strategies to navigate this vital aspect of your cloud transformation journey.

Chapter 7: Data Mastery in the Cloud: Overcoming Migration and Integration Hurdles

Moving your applications to the cloud is like upgrading to a high-speed train - sleek, efficient, and fast. But what good is a shiny new train without tracks to run on? In this case, those tracks are your data - the lifeblood of your organization. Without the right strategy, your journey to the cloud could feel more like a train derailment than a seamless ride. This chapter takes you through the critical aspects of data migration, integration, security, and compliance in the cloud, ensuring your data is not only along for the ride but driving your success.

Data Migration Without Disruption: A Smooth Transition

Migrating data to the cloud can feel like trying to relocate a bustling library—all while patrons are still checking out books. It's not just about packing boxes and hoping for the best; it's about having a detailed plan to keep operations running smoothly. Here's how to make sure your data migration doesn't turn into a chaotic mess of misplaced information and downtime complaints.

Phased Data Migration: The Strategic Approach

Phased data migration is a strategy where you move your data to the cloud in incremental stages, rather than all at once. Think of it as a carefully orchestrated move, where you pack and transport your belongings room by room instead of trying to move the entire house in one go. This controlled approach is often the backbone of a successful cloud migration, especially for complex or large datasets.

Key Characteristics of Phased Migration:

- Incremental Approach: Data is migrated in smaller, manageable chunks, allowing for better control and risk mitigation.

- Prioritization: You prioritize data migration based on factors like business criticality, data sensitivity, and application dependencies.

- Testing and Validation: Each phase includes thorough testing and validation to ensure data integrity and application functionality before moving to the next stage.

- Flexibility: The phased approach allows for adjustments and course correction based on the learnings from each phase.

- Reduced Disruption: By migrating data incrementally, you minimize disruption to business operations and ensure a smoother transition.

Benefits of Phased Data Migration:

- Reduced Risk: Migrating in smaller chunks allows you to identify and address potential issues early on, reducing the risk of a major data migration failure.

- Improved Control: You have greater control over the migration process, allowing for better management of resources, timelines, and budgets.

- Increased Flexibility: You can adjust your migration strategy based on the learnings from each phase, accommodating unforeseen challenges or changing business requirements.

- Minimized Downtime: By migrating data incrementally, you can minimize downtime and ensure business continuity.

- Enhanced Confidence: The phased approach allows you to build confidence in the migration process and ensure a smoother transition to the cloud.

Use Case Example: "Global Retailer"

Scenario: Global Retailer, a multinational retail giant with a vast network of stores and an extensive online presence, wants to migrate its massive product catalog and customer data to the cloud to improve scalability, performance, and business agility.

Challenge: Migrating such a large and complex dataset all at once poses significant risks and could disrupt critical business operations.

Solution: Global Retailer adopts a phased data migration strategy, dividing the migration into four key phases:

- Phase 1: Pilot Migration: They start by migrating a small subset of their product catalog and customer data for a specific region or product category. This allows them to test their migration tools, processes, and cloud environment.

- Phase 2: Customer Data Migration: After a successful pilot, they migrate all customer data to the cloud, prioritizing data related to their loyalty program and online store. This enables them to leverage cloud-based analytics for personalized marketing and customer segmentation.

- Phase 3: Product Catalog Migration: They migrate their entire product catalog to the cloud, ensuring data consistency and accuracy across all sales channels. This allows them to improve product search, recommendations, and inventory management.

- Phase 4: Historical Data Migration: Finally, they migrate historical sales and transactional data to the cloud for long-term storage and analysis. This enables them to gain insights into customer trends and optimize their business strategies.

Benefits:

- Reduced Risk: By migrating in phases, Global Retailer minimizes the risk of data loss or disruption to their online store and retail operations.

- Improved Control: They maintain control over the migration process, ensuring data integrity and validating each phase before proceeding to the next.

- Increased Flexibility: They can adjust their migration strategy based on the learnings from each phase, accommodating any unforeseen challenges.

- Minimized Downtime: They ensure business continuity by migrating data incrementally and avoiding a complete shutdown of their systems.

- Enhanced Confidence: The phased approach builds confidence in the migration process and ensures a smoother transition to the cloud.

Phased data migration is a prudent approach for large and complex data migration projects. It allows organizations like Global Retailer to mitigate risks, maintain control, and ensure a smooth transition to the cloud. By breaking down the migration into manageable stages, they can achieve their cloud objectives without disrupting their business operations.

Data Assessment and Cleansing: The Unsung Hero

Think of your data as the foundation of your business. It's the raw material that fuels your operations, informs your decisions, and drives your growth. But data can also be messy, inconsistent, and riddled with errors. Migrating this messy data to the cloud "as is" is like packing up a dusty, disorganized attic and unpacking it in your new, pristine home. You'll end up with the same clutter, just in a different location.

Data assessment and cleansing address this challenge by:

- Identifying and resolving data quality issues: This includes inconsistencies, errors, duplicates, and missing values that can hinder your business operations and cloud migration efforts.

- Ensuring data integrity: Accurate, complete, and consistent data is crucial for making informed decisions, complying with regulations, and maintaining customer trust.

- Optimizing cloud resources: Clean and well-structured data requires less storage space, improves query performance, and reduces processing costs in the cloud.

- Enabling data integration: Cleansed data is easier to integrate with other systems and applications, breaking down data silos and facilitating a unified view of your business.

- Supporting data analytics and machine learning: High-quality and high-volume data are essential for accurate and reliable data analysis, enabling you to extract meaningful insights and drive business value.

In essence, data assessment and cleansing lay the groundwork for a successful cloud migration and ensure that your data remains a valuable asset in your new cloud environment. To achieve this pristine state, a structured approach is necessary, starting with understanding your data, followed by applying targeted cleansing techniques.

The Data Assessment Process: Knowing Your Data Inside Out

Before you can cleanse your data, you need to understand its current state. Data assessment involves a comprehensive evaluation of your data's quality, structure, and relevance to your business needs. This process typically includes:

- Data discovery and profiling: Identify all your data sources, understand the types of data you have, and analyze its characteristics, such as data volume, data types, and data formats.

- Data quality analysis: Evaluate the accuracy, completeness, consistency, and validity of your data. Identify any errors, inconsistencies, duplicates, or missing values.

- Data mapping and lineage: Trace the flow of data within your organization, understanding how data is collected, processed, stored, and used. This helps identify potential data quality issues and dependencies between different systems.

- Data risk assessment: Identify potential risks associated with your data, such as security vulnerabilities, compliance gaps, or data privacy concerns.

- Data value assessment: Determine the business value of your data and prioritize data cleansing efforts based on its importance to your operations and strategic goals.

This assessment provides a comprehensive understanding of your data landscape, highlighting areas that need attention and guiding your cleansing strategy.

Data Cleansing Techniques: Tidying Up Your Data

Data cleansing involves a variety of techniques to address different types of data quality issues. Here are some common approaches:

- Data deduplication: Identify and remove duplicate records, ensuring each piece of information is stored only once.

- Data standardization: Enforce consistent formats and values for data fields, such as dates, addresses, and names.

- Data validation: Check data against predefined rules or constraints to identify and correct invalid or inaccurate values.

- Data enrichment: Enhance existing data with additional information from external sources (e.g., demographic data or geographic coordinates).

- Data transformation: Convert data from one format to another, such as transforming text data to numerical values or vice versa.

- Missing value imputation: Fill in missing data values using statistical methods or logical inferences.

The specific techniques you use will depend on the nature of your data and the types of data quality issues you encounter.

Tools and Technologies for Data Assessment and Cleansing

A variety of tools and technologies can assist with data assessment and cleansing. These include:

- Data profiling tools: These tools automate the process of analyzing data characteristics, identifying data quality issues, and generating reports.

- Data quality rules engines: These tools allow you to define and enforce data quality rules, automatically identifying and correcting data errors.

- Data cleansing software: These software packages provide a suite of tools for data deduplication, standardization, validation, and enrichment.

- Cloud-based data management services: Cloud providers offer various services for data integration, transformation, and quality management, simplifying the data cleansing process.

Choosing the right tools and technologies can significantly streamline your data assessment and cleansing efforts, saving time and resources.

The Importance of Ongoing Data Quality Management

Data assessment and cleansing are not one-time activities. Data quality is an ongoing concern that requires continuous monitoring and maintenance. This involves:

- Establishing data quality metrics: Define key metrics to track data quality over time, such as the number of duplicates, error rates, or completeness scores.

- Implementing data quality controls: Establish processes and procedures to prevent data errors from entering your systems in the first place.

- Regular data audits: Conduct periodic data audits to identify and address any emerging data quality issues.

- Data governance: Establish clear roles and responsibilities for data management and ensure data quality is a shared responsibility across the organization.

By embedding data quality management into your organizational culture and processes, you can ensure that your data remains a valuable asset throughout your cloud journey and beyond.

Data Replication: Minimizing Downtime

Data replication is the process of creating and maintaining an exact copy of your data in a separate location. Think of it as creating a backup of your important files, but for your entire database or data store. This copy can reside on a different server, in a different data center, or even in a different cloud environment.

Why is Data Replication Important in Data Migration?

Data replication plays a crucial role in data migration projects for several reasons:

- Minimizing Downtime: By continuously replicating data to the cloud before the final cutover, you can significantly reduce downtime during the migration process. This ensures business continuity and avoids disruptions to your operations.

- Testing and Validation: You can use the replicated data in the cloud to test your applications and validate the migration process before making the final switch. This allows you to identify and address any issues before they impact your live environment.

- Data Backup and Recovery: Data replication provides an additional layer of data protection. If something goes wrong with your primary data store, you have a readily available copy in the cloud to restore from.

- Facilitating Phased Migrations: Data replication enables you to migrate data in phases, starting with less critical data or a subset of your overall dataset. This reduces risk and allows for a more controlled migration process.

- Supporting Hybrid Cloud Scenarios: Data replication can be used to keep data synchronized between on-premises and cloud environments, enabling hybrid cloud deployments and ensuring data consistency.

How Does Data Replication Work in a Data Migration Project?

Here's a simplified overview of how data replication is used in a data migration project:

- Initial Replication: You use a data replication tool to create an initial copy of your data in the cloud environment. This might involve a full snapshot of your data or an incremental replication that captures only the changes made since the last replication.

- Continuous Synchronization: The replication tool continuously monitors your source data and replicates any changes to the cloud copy in near real-time. This ensures that the cloud copy remains up-to-date with your primary data store.

- Testing and Validation: You use the replicated data in the cloud to test your applications, validate the migration process, and ensure everything works as expected in the new environment.

- Cutover: Once you're confident in the migrated data and applications, you switch over to the cloud environment. This might involve redirecting traffic to your cloud-based applications and making the cloud copy your primary data store.

- Ongoing Replication (Optional): Depending on your needs, you might continue replicating data between your on-premises and cloud environments to support hybrid cloud scenarios or maintain a backup in the cloud.

Types of Data Replication:

- Synchronous Replication: Changes are replicated to the target environment in real-time, ensuring high consistency but potentially impacting performance.

- Asynchronous Replication: Changes are replicated with a slight delay, offering better performance but with a potential for data loss in case of a failure before replication completes.

- Near Real-time Replication: Changes are replicated with minimal delay, offering a balance between consistency and performance, often preferred for migrations.

Choosing the Right Data Replication Tool:

Several data replication tools are available, each with its own strengths and weaknesses. Major cloud providers (e.g., AWS Database Migration Service, Azure Data Sync, Google Cloud Database Migration Service) offer robust services, but third-party tools are also common. When choosing a tool, consider factors like:

- Supported Data Sources and Targets: Ensure the tool supports your existing databases and the cloud environment you're migrating to.

- Replication Methods: Choose a tool that offers the appropriate replication methods (synchronous, asynchronous, or near real-time) for your needs.

- Performance and Scalability: Select a tool that can handle the volume and frequency of your data changes without impacting performance.

- Ease of Use and Management: Choose a tool that is easy to configure, monitor, and manage.

- Cost: Consider the cost of the tool and any associated licensing fees.

Data Transfer Optimization: Speeding Up the Move

Data transfer optimization is like finding the fastest and most efficient route for your data to travel to its new home in the cloud. It's about minimizing the time, cost, and complexity of moving large datasets, while ensuring data integrity and security. This is particularly crucial for large initial data loads in a phased migration.

Here's a breakdown of what it involves:

1. Choosing the Right Transfer Method:

- Online Transfer: This involves transferring data over the internet using tools like the cloud provider's console, command-line tools, or third-party data migration services. It's suitable for smaller datasets or less time-sensitive migrations.

- Offline Transfer: This involves physically shipping data to the cloud provider using dedicated appliances or storage devices (e.g., AWS Snow Family, Azure Data Box, Google Transfer Appliance). It's often faster and more cost-effective for extremely large datasets, especially when internet bandwidth is limited.

- Direct Connect: This involves establishing a dedicated, private network connection between your on-premises infrastructure and the cloud provider's network. It offers the highest bandwidth and security but can be more expensive and requires a higher upfront investment.

The best method depends on factors like the size of your data, your network bandwidth, your budget, and your time constraints.

2. Optimizing Network Performance:

- Bandwidth Optimization: Maximize your available bandwidth by optimizing network configurations, prioritizing data transfer traffic, and minimizing competing network activities.

- Compression: Compress data before transferring it to reduce the amount of data that needs to be transmitted, improving transfer speeds and reducing costs.

- Data Deduplication: Identify and eliminate duplicate data before transferring it to reduce the overall data volume and improve efficiency.

- WAN Optimization: Utilize WAN optimization techniques to improve data transfer performance over long distances, especially for geographically distributed organizations.

3. Ensuring Data Security During Transfer:

- Encryption: Encrypt data both in transit and at rest to protect it from unauthorized access and ensure data confidentiality.

- Secure Transfer Protocols: Use secure transfer protocols like HTTPS or SFTP to protect data during transmission.

- Access Control: Implement robust access control mechanisms to restrict access to sensitive data during the transfer process.

- Data Integrity Checks: Implement data integrity checks (e.g., checksums) to ensure that data is not corrupted or modified during transfer.

4. Monitoring and Managing the Transfer Process:

- Real-time Monitoring: Monitor the data transfer process in real-time to track progress, identify potential issues, and ensure optimal performance.

- Alerting and Notifications: Set up alerts and notifications to be informed of any errors or delays during the transfer process.

- Logging and Reporting: Maintain detailed logs and generate reports to track the progress and success of your data transfer activities.

Data Validation and Testing: Your Quality Control Checkpoints

Data validation and testing are critical steps in any data migration project, especially when moving to the cloud. They act as your quality control checkpoints, ensuring your data arrives safely and functions correctly in its new environment. Think of it as a thorough inspection before moving into a new home – you want to make sure everything is in working order and that there are no hidden surprises.

Here's a breakdown of what data validation and testing entail:

Data Validation:

What it is: Data validation focuses on checking the accuracy, completeness, and consistency of your data *after* it has been migrated to the cloud. It's like comparing your belongings to an inventory list after the move to ensure everything arrived intact and in the right place.

How it works: You use various techniques and tools to compare the migrated data against the original source data or predefined rules and standards. This might involve:

- Data profiling: Analyzing the characteristics of the migrated data, such as data types, formats, and distributions, to identify any anomalies or inconsistencies.

- Data comparisons: Comparing the migrated data with the original source data to identify any discrepancies or errors.

- Rule-based validation: Checking data against predefined rules or constraints, such as data type checks, range checks, or format validation.

- Cross-field validation: Verifying the consistency of data across different fields, such as ensuring that a customer's address matches their zip code.

Data Testing:

What it is: Data testing goes beyond basic validation to ensure that your data works correctly within your applications and business processes in the cloud environment. It's like testing all the appliances and systems in your new home to make sure they function as expected.

How it works: You perform various tests to verify that your applications can access, process, and utilize the migrated data correctly. This might involve:

- **Functional testing**: Testing the functionality of your applications that rely on the migrated data, such as verifying that customer orders can be processed correctly or that reports generate accurate results.

- **Integration testing**: Testing the integration of your applications with other systems and services in the cloud environment, ensuring data flows seamlessly between different components.

- **Performance testing**: Testing the performance of your applications with the migrated data, ensuring they can handle the expected workload and data volumes.

- **User acceptance testing (UAT)**: Involving end-users in testing the migrated data and applications to ensure they meet business requirements and user expectations.

Why Data Validation and Testing are Important:

- **Ensuring Data Integrity**: They ensure that your data remains accurate, complete, and consistent throughout the migration process.

- **Preventing Costly Errors**: They help identify and correct data errors before they impact your business operations or lead to costly rework.

- **Maintaining Compliance**: They ensure that your data meets regulatory requirements and industry standards.

- **Improving Data Quality**: They help improve the overall quality of your data, enabling better decision-making and business insights.

- **Building Confidence**: They provide confidence that your data is migrated correctly and that your applications will function as expected in the cloud.

By investing time and resources in these activities, leveraging automated tools wherever possible, you can minimize risks, avoid costly errors, and ensure a smooth and successful transition to the cloud.

Unifying Disparate Systems: Breaking Down Data Silos

Legacy systems can be like cliques at a high school football game—refusing to mingle and keeping their valuable intel to themselves. Cloud solutions are the ultimate icebreaker, unifying your data and turning isolated systems into one big, happy, communicative family.

Achieving this unification requires strategic approaches and the right tools.

Key Tools and Approaches for Data Integration

Data integration is about making different systems "talk" to each other, consolidating data into a central repository, and breaking down walls that keep valuable insights locked away.

- Data Warehousing: Once you've unified your data, you need a place to store it. Cloud-based data warehouses are like climate-controlled vaults—scalable, secure, and perfect for storing large volumes of structured data from diverse sources for analytical purposes.

- Data Lakes: If a data warehouse is a meticulously organized library, a data lake is like a giant, flexible sandbox. Whether it's structured, semi-structured, or unstructured data, a cloud-based data lake lets you store, explore, and even build predictive models with ease, offering flexibility for future uses.

- API Gateways: Think of these as secure, guarded gates allowing applications to share data without spilling secrets. They ensure that only authorized personnel or systems have access to sensitive information, providing a controlled and secure way for disparate systems to communicate.

Data Integration with iPaaS

In today's interconnected world, data is often scattered across various applications, systems, and databases, creating data silos that hinder an organization's ability to gain a complete view of its operations and make informed decisions. iPaaS (Integration Platform as a Service) solutions have emerged as a powerful tool to address this challenge, particularly in the context of cloud migration and ongoing integration.

What is iPaaS?

iPaaS is a cloud-based platform that enables organizations to connect different applications and data sources, regardless of their location (on-premises, cloud, or hybrid). It provides a centralized environment for data integration, allowing organizations to streamline data flow, automate processes, and gain a unified view of their data.

Benefits of iPaaS:

- Simplified Integration: iPaaS solutions offer pre-built connectors and templates that simplify the integration process, reducing the need for complex coding and development.

- Improved Data Flow: By connecting disparate systems, iPaaS enables seamless data flow across the organization, breaking down data silos and facilitating better collaboration.

- Reduced Development Time: iPaaS solutions accelerate integration projects by providing a visual development environment and pre-built components, freeing up IT resources for other strategic initiatives.

- Increased Agility: iPaaS allows organizations to quickly adapt to changing business needs by easily adding or modifying integrations as required.

Boomi and MuleSoft: Leading iPaaS Examples

Two leading iPaaS solutions are Boomi and MuleSoft:

- Boomi: Boomi is a cloud-native iPaaS platform that offers a wide range of connectors, a user-friendly interface, and robust data mapping capabilities. It is known for its ease of use and scalability, making it suitable for organizations of all sizes.

- MuleSoft: MuleSoft's Anypoint Platform is a comprehensive iPaaS solution that provides a unified platform for API management, data integration, and application development. It is known for its flexibility and enterprise-grade features, making it a popular choice for large organizations with complex integration needs.

Master Data Management (MDM)

Master Data Management (MDM) is a critical component of data integration, ensuring that the data you're connecting is consistent, accurate, and reliable. MDM focuses on creating a single, "golden record" of key data entities, like customers, products, or employees, across an organization. It's the process of defining, managing, and maintaining consistent, accurate, and complete master data for an organization.

Benefits of MDM:

- Improved Data Quality: MDM helps to cleanse, standardize, and deduplicate data, leading to higher quality data for analysis and decision-making.

- Reduced Data Redundancy: By creating a single source of truth for master data, MDM eliminates data inconsistencies and redundancies across different systems.

- Better Decision-Making: Accurate and consistent master data provides a reliable foundation for business intelligence and analytics, leading to better insights and more informed decisions.

- Increased Efficiency: MDM streamlines data management processes, reducing manual effort and improving operational efficiency.

MDM in the Cloud Context: Cloud environments often consolidate data from various sources, making MDM even more crucial to prevent new data silos or inconsistencies from forming. Cloud-based MDM solutions and platforms that integrate MDM capabilities (like Boomi) simplify the deployment and management of master data.

Use Case Example (Integrated iPaaS & MDM):

A healthcare provider used Boomi's platform to integrate data from various clinical systems, patient portals, and administrative databases. Using Boomi's iPaaS capabilities, they connected these systems to create a unified view of patient data. Additionally, they leveraged Boomi's MDM features to ensure that patient data was consistent, accurate, and up-to-date across all systems. This improved patient care, streamlined operations, and ensured compliance with healthcare regulations.

Unified data isn't just about tearing down walls; it's about building bridges to better decision-making and deeper insights.

Ensuring Data Security and Compliance: Protecting Your Assets
When it comes to your data, security isn't just a line item—it's the foundation of trust. In the cloud, security and compliance aren't optional; they're mission-critical. Your data deserves a digital fortress, not a flimsy lock on the front door. Here's how to protect your most valuable assets, understanding that security in the cloud is a shared responsibility between you and your cloud provider.

- Data Encryption: Data encryption is like putting your valuables in a safe. Whether in transit or at rest, your data should be encrypted to ensure that even if someone intercepts it, all they get is digital gibberish.

- Access Control: Remember the bouncer at your favorite nightclub? That's access control for your cloud environment. Only authorized personnel should have access, and their permissions should match their job responsibilities—no sneaking into the VIP section uninvited. Implement principles like least privilege (granting only necessary access) and role-based access control.

- Compliance Certifications: Not all cloud providers are created equal. Look for providers with certifications that align with your industry's regulations, whether that's HIPAA for healthcare, PCI DSS for payments, or ISO 27001 for information security.

- Data Residency: Just as certain wines can only be made in specific regions, some data must remain in designated geographic locations due to legal or regulatory requirements (e.g., GDPR mandates for EU citizen data). Ensure your cloud provider complies with data residency laws to avoid regulatory headaches.

- Security Monitoring and Auditing: Think of this as hiring a 24/7 security team. Continuous monitoring and regular audits can detect threats before they escalate, keeping your cloud environment safe. This includes logging, anomaly detection, and security information and event management (SIEM) systems.

A strong security and compliance strategy doesn't just protect your data—it protects your reputation, your customers, and your bottom line.

Conclusion

Addressing data challenges in the cloud isn't just a technical exercise— it's a critical component of your digital transformation strategy. With a well-defined plan for data migration, powerful tools to unify disparate systems, and robust security and compliance measures, you'll do more than move your data to the cloud—you'll unlock its full potential.

Whether you're keeping operations running during migration, breaking down stubborn data silos, or building a digital fortress of security, your efforts will pay dividends in better insights, smoother operations, and long-term business success. By embracing these strategies, your organization can truly harness the power of cloud data, driving innovation and achieving sustainable growth.

Chapter 8: Bridging the IT and Business Divide: A Collaborative Approach to Legacy System Transformation

The journey of modernizing legacy systems with cloud solutions is not just a technological endeavor; it's a transformative process that requires a deep understanding of both the intricate workings of IT infrastructure and the strategic objectives of the business. Indeed, studies show that a significant percentage of technology transformations - some estimates suggest over 70% - falter or outright fail, not due to technical hurdles, but because of a critical **disconnect between IT and business priorities**. This chapter delves into the essential human element of this transformation, emphasizing the importance of collaboration, communication, and a shared vision across the organization.

We'll explore how to build bridges between the IT department and business units, fostering a culture of innovation and shared success. Along the way, we'll introduce new approaches, such as embedding agile methodologies into cross-functional teams and leveraging organizational change management (OCM) principles to ease transitions

The Role of Shared Objectives: Setting the Foundation for Success

Legacy system transformations often fail because IT and business leaders operate with misaligned priorities. Bridging the divide begins with defining shared objectives that reflect both technological imperatives and business goals.

A shared objective could be as straightforward as, "Deliver a seamless customer experience through real-time data insights," or as ambitious as, "Enable full digital transformation to pivot toward a subscription-based revenue model."

By establishing these objectives collaboratively:

- **IT gains context** Understanding the business drivers behind modernization efforts allows IT to prioritize initiatives that align with growth and efficiency goals.
- **Business leaders engage**: Seeing the technical roadmap in the context of ROI and competitive advantage builds executive buy-in.
- **Stakeholders unify**: Shared goals serve as a rallying point, helping diverse teams collaborate effectively.

The key is to tie every technical milestone back to a measurable business impact.

Creating Cross-Functional Teams: Aligning IT, Business Units, and Stakeholders

Legacy systems often operate in silos, reflecting a time when technology was viewed as a separate entity from the core business functions. However, in today's interconnected world, technology is the backbone of every successful business. To effectively leverage cloud solutions for legacy system enhancement, it is crucial to break down these silos and create cross-functional teams that align IT, business units, and key stakeholders.

These teams should be composed of individuals with diverse skill sets and perspectives, including:

- **IT specialists:** These individuals possess technical expertise to assess the current legacy system, identify areas for improvement, and recommend appropriate cloud solutions. They understand the intricacies of cloud migration, security protocols, and integration challenges.

- **Business analysts**: Representing the voice of the business, these team members articulate the specific needs and challenges faced by different departments. They translate business requirements into technical specifications, ensuring that the cloud solution addresses real-world problems and contributes to strategic goals.
- **End-users**: Incorporating end-users from various departments provides invaluable insights into the daily operational challenges and desired improvements. Their feedback ensures that the new system is user-friendly, efficient, and aligned with their workflow.
- **Project managers**: These individuals oversee the entire modernization project, ensuring that it stays on track, within budget, and meets its objectives. They facilitate communication and collaboration among team members, mitigating risks and resolving conflicts.
- **Executive sponsors**: Having a member of the leadership team to champion the project provides high-level support and ensures alignment with the overall business strategy. Their involvement helps to secure resources, remove roadblocks, and communicate the project's value to the wider organization.

By bringing together these diverse perspectives, cross-functional teams can:

- **Develop a shared understanding of the project goals**: This ensures that everyone is working towards the same objectives and that the chosen cloud solution addresses the needs of all stakeholders.
- **Identify potential challenges and risks early on**: This allows for proactive mitigation strategies and reduces the likelihood of costly delays or failures.
- **Foster a sense of ownership and accountability**: When individuals from different departments are involved in the decision-making process, they are more likely to be invested in the project's success.

- **Promote innovation and creativity**: By bringing together different perspectives and expertise, cross-functional teams can generate more innovative solutions and identify new opportunities for leveraging cloud technology.

Integrating Agile Principles for Collaboration

Beyond simply forming cross-functional teams, truly **embedding agile methodologies** into their day-to-day operations can dramatically accelerate the bridging of the IT and business divide. Agile isn't just for software development; its core principles inherently promote the very collaboration this chapter champions. By embracing agile, teams can:

- **Foster Continuous Feedback Loops:** Agile frameworks, through ceremonies like daily stand-ups, sprint reviews, and retrospectives, mandate constant interaction. This ensures business stakeholders are regularly providing input and seeing progress, while IT gains immediate clarity on evolving business needs and challenges. No more waiting months for feedback; issues are identified and addressed in real-time.

- **Prioritize Business Value:** Agile rituals, particularly sprint planning and backlog refinement, force joint prioritization. Business units and IT must collectively decide what delivers the most immediate value, ensuring that technical efforts directly align with strategic business goals and customer needs. This transparent prioritization builds shared ownership over the product roadmap.

- **Encourage Iterative Delivery and Learning:** Instead of monolithic, long-term projects, agile promotes delivering functionality in small, usable increments. This iterative approach allows business users to start seeing tangible benefits sooner, test new features, and provide feedback that shapes subsequent iterations. It builds trust by demonstrating continuous progress and adaptability, reducing the "big bang" risk often associated with legacy transformations.

- **Promote Self-Organizing Teams:** Agile empowers cross-functional teams to make decisions about *how* they will achieve their goals. This decentralization fosters a sense of ownership and accountability within the team, encouraging innovation and problem-solving at the ground level, often leading to more creative and efficient solutions for business challenges.

By adopting agile, organizations create a rhythm of collaboration that transforms IT-business interactions from sequential hand-offs into a dynamic, continuous partnership, making the modernization journey far more adaptive and successful.

Communicating Value: Framing Technical Transformation in Terms of Business Outcomes

While IT professionals may be excited about the technical intricacies of cloud migration and the potential for enhanced functionality, it's essential to communicate the value of these changes in a language that resonates with business stakeholders. Instead of focusing solely on technical jargon and specifications, frame the transformation in terms of tangible business outcomes.

Here are some strategies to effectively communicate the value of cloud-enhanced legacy systems:

- **Quantify the benefits**: Whenever possible, use data and metrics to demonstrate the potential return on investment (ROI). This could include cost savings from reduced infrastructure maintenance, increased efficiency through automation, or revenue growth from improved customer experience.
- **Focus on solving business problems**: Clearly articulate how the cloud solution will address specific pain points experienced by different departments. This could include improving data accessibility, streamlining workflows, or enhancing collaboration across teams.

- **Use storytelling and case studies**: Illustrate the benefits of cloud transformation with real-world examples and success stories from other organizations. This helps to make the concept more tangible and relatable for business stakeholders.
- **Highlight the competitive advantage**: Explain how modernizing legacy systems with cloud solutions can give the organization a competitive edge in the market. This could include increased agility, faster time-to-market, or improved customer satisfaction.
- **Emphasize resilience**: Highlight how cloud solutions safeguard business continuity against disruptions like cyberattacks or market shifts.
- **Champion employee empowerment**: Showcase how modern tools simplify workloads, enabling employees to focus on strategic, high-value activities.
- **Inspire growth opportunities**: Explain how legacy modernization opens doors to innovation, such as adopting AI-driven insights or entering new markets.
- **Maintain open and transparent communication**: Regularly update stakeholders on the project's progress, challenges, and successes. This fosters trust and ensures that everyone is informed and aligned.

By effectively communicating the value of cloud transformation in business terms, IT leaders can gain buy-in from stakeholders, secure necessary resources, and ensure the project's success.

Driving Engagement Through Organizational Change Management (OCM)

While technology forms the backbone of legacy system modernization, it's the people within the organization who ultimately determine its success. Introducing new cloud solutions and transforming established processes can be disruptive, often leading to resistance and hindering adoption if not managed carefully. This is where Organizational Change Management (OCM) becomes crucial. OCM is a structured approach to guiding individuals, teams, and the organization through a transition to a desired future state. Incorporating OCM strategies into your legacy system modernization project is essential to ease these transitions by proactively addressing potential resistance and fostering widespread adoption of the new systems and processes.

Here are key OCM strategies to consider:

- **Stakeholder Analysis**: The first step in any OCM effort is to thoroughly understand the stakeholders involved. This involves identifying all individuals and groups who will be affected by the change, assessing their level of influence, and understanding their potential concerns or hesitations. A stakeholder analysis helps tailor communication and support strategies to specific groups, ensuring that their needs are addressed and their concerns are acknowledged. By understanding where resistance might arise, you can proactively address it.

- **Change Champions**: Enlisting change champions is a powerful way to promote adoption from within. These are influential employees from various departments who are enthusiastic about the project and willing to advocate for it among their peers. Change champions play a vital role in communicating the benefits of the new system, answering questions, addressing concerns, and providing support during the transition. They act as liaisons between the project team and their respective departments, fostering trust and facilitating buy-in.

- **Training Programs**: Adequate training is essential to ensure that employees feel confident and competent in using the new cloud-based systems. Training programs should be tailored to different roles and skill levels, providing hands-on experience and practical guidance. In addition to initial training, ongoing support and resources should be available to reinforce learning and address any questions that may arise. Investing in comprehensive training programs demonstrates a commitment to employee success and empowers them to embrace the new technology.

- **Communication Plans**: Transparent and consistent communication is paramount throughout the entire modernization journey. A well-defined communication plan should outline how information will be shared, when, and with whom. Regular updates, progress reports, and feedback sessions help to keep stakeholders informed, manage expectations, and build trust. Communication should be two-way, encouraging feedback and addressing concerns promptly.
- **Feedback Loops**: Establishing mechanisms for continuous feedback is crucial for adapting the implementation approach and addressing unforeseen challenges. This can involve creating forums for employees to share their experiences, providing surveys to gather input, and actively soliciting suggestions for improvement. By actively listening to feedback and responding accordingly, the project team can demonstrate responsiveness and ensure that the new system meets the needs of its users.

In essence, OCM is not merely a soft skill; it's a critical discipline that complements the technical aspects of legacy system modernization. By prioritizing the human side of change, organizations can build trust, minimize disruption, and maximize the likelihood of successful adoption and long-term value creation.

Leveraging Data as a Unifier

Data has the remarkable potential to act as a common ground, bridging the gap between IT and business units. In many organizations, IT departments manage the systems that generate and store data, while business units rely on that data to make informed decisions. However, legacy systems often create data silos, making it difficult for both IT and business users to access and utilize information effectively. Modernizing these systems with cloud solutions presents a valuable opportunity to break down these silos and democratize data, enabling both departments to collaborate more effectively on shared insights and drive business value.

Here's how to leverage data effectively as a unifier:

- **Create a centralized data hub**: Establishing a single source of truth is paramount. This involves building a centralized data hub or data warehouse that consolidates data from various legacy systems and makes it accessible to both IT and business users. This eliminates the confusion and inconsistencies that arise from having multiple versions of the truth and ensures that everyone is working with the same reliable information. A centralized data hub fosters transparency, improves data quality, and facilitates collaboration by providing a shared foundation for analysis and decision-making.

- **Foster data literacy**: Democratizing data goes beyond simply providing access; it also requires empowering business users to understand and utilize data effectively. This involves investing in data literacy programs and workshops that equip business users with the skills and knowledge to interpret data, perform basic analysis, and use data visualization tools. By fostering data literacy, organizations enable business users to become more self-sufficient in their data needs, reducing their reliance on IT for every query and report. This not only improves efficiency but also empowers business users to make more data-driven decisions.

- **Develop KPIs together**: Key Performance Indicators (KPIs) are crucial metrics that track progress towards specific goals. When IT and business leaders collaborate on developing KPIs, they ensure that these metrics reflect shared priorities and align with overall business objectives. This collaborative process fosters a sense of ownership and accountability, as both departments have a vested interest in achieving the targets. Furthermore, it promotes a common understanding of what constitutes success and facilitates meaningful dialogue around performance and areas for improvement. Examples of shared KPIs could include customer retention rates, operational efficiency gains, or revenue growth attributed to specific IT initiatives.

In conclusion, data serves as a powerful tool for fostering collaboration and alignment between IT and business units. When IT and business leaders work together on data initiatives, they not only establish trust but also unlock the full potential of data to drive informed decision-making, improve business outcomes, and achieve mutual wins.

Empowering Collaboration Through Digital Collaboration Tools

In today's dynamic work environment, characterized by hybrid and remote work models, digital collaboration tools have become indispensable for breaking down traditional barriers between departments and fostering seamless teamwork. These tools provide the means for teams to connect, communicate, and collaborate effectively, regardless of physical location. By facilitating real-time interaction, streamlining workflows, and enhancing transparency, digital collaboration tools play a crucial role in bridging the IT and business divide and driving successful legacy system transformations.

Here's a deeper look at how these tools empower collaboration:

Real-time Communication and Brainstorming: Platforms like Slack or Microsoft Teams provide channels for instant messaging, group discussions, and file sharing, enabling real-time communication and fostering a sense of community across departments. These tools facilitate quick problem-solving, spontaneous brainstorming sessions, and informal interactions, which are essential for building relationships and fostering a collaborative spirit. The ability to create dedicated channels for specific projects or teams ensures that relevant information is easily accessible and that communication remains focused.

Project Management and Workflow Visualization: Tools such as Asana or Jira enable teams to track project milestones, assign tasks, visualize dependencies, and manage workflows in a transparent and organized manner. These platforms provide a central hub for project-related information, ensuring that everyone is on the same page and that progress is easily monitored. By streamlining task management and improving accountability, these tools enhance efficiency and reduce the risk of miscommunication or delays.

Data Visualization and Shared Insights: Platforms like Power BI or Tableau empower teams to collaboratively build and share interactive dashboards that connect technical achievements to key business metrics. These tools transform raw data into meaningful visualizations, making it easier for both IT and business users to understand the impact of technology initiatives on business outcomes. By fostering data transparency and enabling shared insights, these tools facilitate data-driven decision-making and promote alignment between IT and business goals.

By seamlessly integrating these digital collaboration tools into daily workflows, organizations can empower their employees to communicate more effectively, collaborate more efficiently, and work together more strategically. This leads to improved communication, increased accountability, and a stronger sense of shared purpose, all of which are essential for bridging the IT and business divide and achieving successful legacy system transformations.

Empowering Collaboration: Fostering a Culture that Supports Innovation

Successfully bridging the IT and business divide requires more than just forming cross-functional teams and communicating effectively. It necessitates fostering a culture that values collaboration, embraces innovation, and empowers individuals to contribute their unique skills and perspectives.

Here are some key elements of a collaborative and innovative culture:

- **Open communication channels**: Encourage open and honest communication across all levels of the organization. This can be achieved through regular meetings, online forums, and feedback mechanisms.
- **Shared vision and goals**: Ensure that everyone understands the organization's strategic objectives and how the cloud transformation project contributes to those goals.
- **Trust and respect**: Create a safe environment where individuals feel comfortable sharing ideas, challenging assumptions, and taking risks.
- **Continuous learning and development**: Provide opportunities for employees to develop their skills and knowledge in cloud technologies and business processes.
- **Recognition and rewards**: Acknowledge and celebrate individual and team contributions to the project's success.

By cultivating a culture that supports collaboration and innovation, organizations can unlock the full potential of cloud solutions to enhance legacy systems and drive business growth. This involves empowering employees to think creatively, challenge the status quo, and embrace new technologies.

VIGNETTE: What Collaboration Looks Like

At Allied Precision Manufacturing, a stalled initiative to improve customer quoting had bounced between sales and IT for months. Sales claimed the system was too rigid; IT insisted it met the documented requirements. Frustration simmered.

That changed when a cross-functional working session was launched. Sales reps, quoting clerks, and IT devs sat around the same table - literally. Within two hours, the root issue was exposed: the quoting logic worked, but the UI flow buried key pricing data three screens deep. A developer created a prototype on the spot.

By the end of the week, a revised interface was live in UAT. No change order. No escalation. Just people, working like a single team.

That's what collaboration looks like.

QUOTE:

"We didn't need a bigger tech stack - we needed to listen to each other. Once marketing and IT started prototyping together instead of passing specs like a hot potato, our product roadmap finally felt like progress, not politics."

- Diane Varela, CEO, Allied Precision Manufacturing

Conclusion

Bridging the IT-business divide isn't about forcing two distinct departments to coexist - it's about transforming them into a dynamic partnership. By forming cross-functional teams, embracing OCM strategies, and communicating value in business terms, organizations can align their technological advancements with strategic goals.

More than just modernizing legacy systems, this collaborative approach unlocks new opportunities for innovation, efficiency, and growth. And perhaps most importantly, it turns the daunting challenge of cloud transformation into a shared journey of success.

Chapter 9: Overcoming Common Barriers

Transforming legacy systems with cloud solutions is not just a technical endeavor - it's a bold leap into the unknown. As thrilling as it sounds, the road is often riddled with obstacles that can trip even the best-laid plans. The barriers may seem insurmountable at first, but they're really just challenges waiting to be solved. This chapter dives into three of the most common hurdles - cultural resistance, budget constraints, and skill gaps - while introducing creative ways to tackle them head-on.

Cultural Resistance: Winning Hearts and Minds

Let's face it: people are creatures of habit. Change - even positive change - can feel like swapping a beloved, beat-up recliner for a sleek, unfamiliar chair that "fits ergonomically." Sure, it's better in every measurable way, but it doesn't have the same cozy charm. This resistance to change is amplified in organizations where processes and systems are deeply entrenched.

Beyond the surface-level concerns of employees clinging to the status quo lies something deeper: a fear of being left behind. It's not just about learning new systems; it's about protecting their sense of purpose and place in the organization. For example: Janet from Accounts Receivable initially resisted the new cloud dashboard. But once she realized she could generate reports in two clicks instead of waiting days for IT, she became its loudest cheerleader.

Breaking the Cycle of Resistance

Here's the thing: resistance isn't the enemy. It's a signal - a call to action that organizations need to address thoughtfully. To win hearts and minds, consider these expanded strategies:

Humanize the change: Move away from corporate jargon. If you're presenting cloud migration to employees, frame it in terms they care about. For example, instead of "streamlined workflows," talk about how they'll spend less time wrestling with unresponsive tools and more time solving interesting problems.

Make leadership relatable: Often, resistance stems from employees feeling that decisions are being made by disconnected executives. Bring leaders into the conversation - not to dictate, but to listen. Stories from leadership about their own struggles with adopting new technologies can create a sense of shared journey.

Establish peer mentors: Employees are more likely to trust peers who've been through similar transitions. Build a network of cloud champions across departments who can act as the first line of support and advocacy.

Normalize learning curves: Acknowledge that mistakes will happen during the transition and create an environment where these missteps are seen as opportunities to learn, not failures to avoid. This shifts the cultural narrative from fear to experimentation.

Going Beyond Engagement

When people resist change, it's often because they can't see themselves in the picture of the future. The real work lies in showing employees their place in the new order and how the cloud isn't a threat to their value but an amplifier of their strengths.

Budget Constraints: Finding Cost-Effective Solutions

Let's talk about the elephant in the room: money. Cloud adoption isn't cheap, especially when you're staring at upfront investments for migration, training, and new tools. It's like deciding to renovate your kitchen - you know it'll pay off with increased functionality and home value, but the initial sticker shock can make you question everything.

Rethinking the ROI Conversation

Too often, organizations look at cloud adoption purely through the lens of cost. Instead, flip the narrative to focus on value creation:

Quantify business outcomes: Don't just say, "This will save money." Highlight the new revenue opportunities cloud capabilities unlock, like faster product launches, improved customer experiences, or better market responsiveness.

Adopt outcome-based pricing models: When engaging with cloud vendors or consultants, negotiate contracts tied to outcomes rather than flat fees. For example, pay a premium for faster deployments that meet specific business goals.

Measure the cost of doing nothing: Legacy systems have hidden costs, from downtime to missed opportunities. Quantify these costs and stack them against the projected cloud investment to build urgency.

Phased Implementation on a Budget

Rome wasn't built in a day, and your cloud infrastructure doesn't have to be either. Consider this phased approach:

- Pilot projects with clear wins: Start with non-critical systems to test the waters and demonstrate quick value.
- Hybrid strategies: Maintain critical legacy systems alongside cloud adoption for a gradual transition.
- Resource-sharing initiatives: Partner with other departments or even external organizations to co-invest in shared cloud services for mutual benefit.

Broader Cost Management Techniques

Beyond the obvious strategies like auto-scaling and reserved instances, think outside the box:

- Invest in "FinOps" teams: These are financial operations teams that specialize in tracking, optimizing, and forecasting cloud spend.
- Build accountability dashboards: Give teams visibility into their cloud usage and costs, encouraging responsible consumption.

Skill Gaps: Upskilling Teams and Leveraging External Expertise

Ah, the skills conundrum. Moving to the cloud often feels like asking a carpenter to build a spaceship - same basic principles, but a wildly different set of tools and techniques. Bridging this gap requires a mix of immediate solutions and long-term strategies.

Short-Term Solutions: Borrowing Expertise

When you need to hit the ground running, external resources can be a lifesaver. But be strategic:

- Adopt a co-delivery model: Work alongside external consultants rather than outsourcing entirely. This ensures your internal teams are learning on the job.
- Target niche expertise: Don't just hire generalists. Look for specialists in areas like cloud-native security or multi-cloud architecture who can provide depth.

Long-Term Solutions: Building an Internal Talent Pipeline

Sustainable cloud adoption requires homegrown expertise. Here's how to get there:

- Embrace role evolution: Shift traditional IT roles toward hybrid roles that combine legacy expertise with cloud fluency.
- Incentivize certifications: Tie skill development to career advancement, offering bonuses or promotions for employees who achieve cloud certifications.

- Rotate roles: Provide cross-functional assignments that expose employees to different cloud disciplines.

The Human Side of Learning

Skills aren't just technical. They're also about adaptability, collaboration, and curiosity. Reinforce that emotional intelligence, curiosity, and collaboration are just as critical as technical fluency. The best cloud teams aren't just certified - they're coachable, adaptable, and hungry to learn. Equip employees with the tools they need to manage not just the technology but the change itself.

Insert: Interconnectedness of Barriers

It's crucial to understand that these barriers - cultural, financial, and technical—rarely exist in isolation. They are deeply **interconnected**, forming a complex web of challenges. For instance, strong cultural resistance can lead to the underutilization of new cloud solutions, effectively wasting precious budget on technology that isn't fully embraced. Similarly, significant skill gaps can prolong migration timelines, escalating costs, and simultaneously fuel employee resistance as they fear being unable to adapt. Conversely, severe budget constraints might limit essential training programs, thereby widening skill gaps and increasing the likelihood of cultural pushback. Recognizing this dynamic interplay is the first step toward crafting truly resilient and comprehensive transformation strategies.

Adding the Bigger Picture: Leadership's Role in Breaking Barriers

Let's zoom out for a moment. These barriers—cultural, financial, and technical—don't exist in silos. They intersect, creating a web of challenges that leadership must navigate. Here's how leaders can rise to the occasion:

- Model resilience: Leaders should embrace cloud adoption as a journey, not a destination. Their attitude sets the tone for the entire organization.

- Foster cross-departmental collaboration: Siloed teams are a recipe for resistance. Leaders should prioritize collaboration, breaking down barriers between IT, finance, HR, and operations.
- Measure success holistically: Track not just technical milestones but also employee engagement, cost alignment, and skill progression. This ensures the transformation is delivering value across the board.

When leaders share both wins and growing pains publicly, it signals that progress - not perfection - is the goal.

Closing Thoughts

Overcoming the barriers to cloud adoption is less about brute force and more about finesse. Think of it as a dance: each step - addressing cultural resistance, managing budgets, and upskilling teams - brings you closer to the rhythm of transformation.

Cloud transformation is ultimately about people, not just technology. By focusing on collaboration, creativity, and communication, you'll not only break through the barriers but pave the way for sustainable success.

And remember: it's not just about surviving the journey. It's about thriving in the destination.

Chapter 10: Scaling Transformation - From Pilot Project to Enterprise-Wide Revolution

Imagine a snowball rolling downhill. It starts small, gains momentum, and becomes an unstoppable force. That's the essence of scaling your cloud journey—moving beyond pilot projects and isolated wins to enterprise-wide adoption that aligns with strategic goals and delivers lasting transformation. This chapter explores key strategies to help you achieve that leap.

Expanding Cloud Capabilities: Beyond Initial Wins

Early modernization successes build confidence, prove value, and generate buy-in. But they're only the beginning. The real impact comes when you scale those wins across your organization, creating a connected, high-performance cloud ecosystem.

Let's say you've migrated a non-critical app to the cloud. You've seen efficiency gains and cost savings. Now what? The goal is to extend those benefits enterprise-wide. Ask yourself:

- How do we apply these lessons across departments?
- How do we integrate services to form a scalable digital backbone?

Key considerations for scaling:

1. Build a Comprehensive Cloud Strategy

Scaling begins with clarity. Your strategy must go beyond cloud adoption—it should clearly define how cloud technology advances business objectives. Include:

- Cloud Adoption Model: Are you going cloud-first or cloud-native? The choice affects scalability and design patterns.
- Business Alignment: Link initiatives to specific business outcomes like agility, cost optimization, or new revenue streams.
- Organizational Readiness:
  - Skills: Do you have the in-house talent, or is upskilling required?
  - Culture: Is the organization open to change and collaboration?
  - Processes: Are your IT workflows cloud-ready?
- Governance Framework: Define policies around cost, security, compliance, and data management to scale safely and sustainably.

2. Prioritize High-Impact Applications

Target workloads that deliver outsized business value with minimal disruption. Don't chase complexity - chase ROI.

3. Adopt a Phased Approach

Instead of a high-risk "big bang" deployment, break your transformation into digestible phases:

Phase 1: Foundation

- Migrate non-critical apps
- Establish core infrastructure and governance
- Launch a Cloud Center of Excellence (CCoE)
- Set up IAM and basic security
- Deploy cost monitoring

Phase 2: Expansion

- Migrate mission-critical workloads
- Integrate data, AI/ML, and automation
- Build CI/CD pipelines and IaC

- Scale compliance and security maturity

Phase 3: Optimization

- Tune cloud spend and performance
- Implement observability and reliability tools
- Leverage serverless and cloud-native architectures
- Instill a culture of continuous innovation

4. Drive Cross-Functional Collaboration

- Cloud isn't just IT's job. Align business units, finance, legal, and ops early and often.

5. Define Success Metrics

- Track outcomes like cost savings, performance improvement, and user satisfaction. Metrics drive momentum.

6. Commit to Continuous Improvement

- Cloud transformation is not a destination—it's a cycle of evolution.

Virtual-World Use Case: Aurevance's Expansion from Pilot to Enterprise Cloud Maturity

The Company (Virtual-World Example)

Aurevance is a fictional mid-sized life sciences company specializing in therapies for rare diseases. Their vast research data and clinical trial information held tremendous potential, but they struggled to extract meaningful insights using legacy on-premises infrastructure.

The Challenge

Aurevance sought to accelerate drug discovery, improve patient outcomes, and enable predictive analytics - but limited compute power, storage capacity, and analytics capabilities created bottlenecks.

The Solution

Recognizing the need for scale, Aurevance partnered with Google Cloud to build a comprehensive data analytics and AI platform using a phased strategy:

Phase 1: Data Migration and Foundation

- o   Migrated research, clinical, and patient data securely
- o   Built a centralized data lake using Google Cloud Storage
- o   Implemented BigQuery for fast and scalable data analysis

Phase 2: Analytics and AI Enablement

- o   Leveraged Looker for data exploration and visualization
- o   Developed AI models for drug discovery using AutoML
- o   Built AI-powered applications supporting personalized medicine

Phase 3: Expansion and Innovation

- o   Integrated genomics pipelines via Google Cloud Genomics
- o   Incorporated real-world data from wearables and patient registries
- o   Enabled collaboration across institutions with Google Cloud's secure sharing capabilities

Strategic Outcomes

This wasn't just a data upgrade - it was a business catalyst. Aurevance accelerated drug development timelines, optimized clinical outcomes, and reduced operational overhead. More importantly, the transformation boosted enterprise value and sharpened their competitive edge.

Much like a snowball gathering speed and mass, Aurevance's measured cloud expansion turned into an unstoppable force for innovation and growth.

Scaling Pitfall to Avoid: Governance Can't Be an Afterthought

Growth without guardrails invites trouble. As you scale:

Risks:

- o Security gaps: Larger attack surface without strong security measures
- o Compliance failures: Especially with HIPAA, GDPR, etc.
- o Cost overruns: Without governance, cloud spend balloons
- o Data silos: Disconnected environments fragment data access

Mitigation Strategies:

- o Governance Framework: Cost control, security, IAM, and resource oversight
- o Security Best Practices:
- o Zero trust model
- o Encryption in transit and at rest
- o Automated threat detection and response

FinOps Discipline:

- o Tag resources
- o Rightsize environments
- o Allocate costs to business units

Data Governance:

- o Define access, quality, and lineage protocols

Building for Scalability

Cloud elasticity is a superpower - but scaling well requires design discipline.

Key Principles:

- o Design for elasticity
- o Embrace automation (IaC, CI/CD, monitoring)
- o Choose scalable services intentionally

- o Continuously monitor and optimize
- o Plan for resiliency and DR

Managing Organizational Change

Cloud at scale isn't just technical - it's cultural. To lead change effectively:

- o Secure executive sponsorship
- o Communicate goals and benefits frequently
- o Invest in training
- o Foster a cloud-first, agile culture

Deploy a change management plan:

- o Stakeholder analysis
- o Communication strategy
- o Training rollout
- o Resistance management
- o Reinforcement mechanisms

Future-Proofing Your Systems

Cloud innovation moves fast - so should you.

Tactics to stay ahead:

- o Track emerging trends (e.g., serverless, edge, AI, quantum)
- o Build modular architectures that flex with change
- o Support continuous learning
- o Create safe zones for experimentation
- o Partner with future-ready providers

Future-proofing isn't about predicting every change - it's about building the agility to adapt when it comes.

Final Thought

Scaling your cloud transformation isn't just about growing your infrastructure - it's about building momentum. When done right, it becomes a force multiplier: accelerating innovation, sharpening your competitive edge, and transforming your business from the inside out.

Chapter 11: Measuring Success: The Key to Proving and Improving Cloud Transformation

Migrating to the cloud is a transformative journey, not merely a single event. While the initial "go-live" marks a significant milestone, the true value of cloud adoption unfolds progressively. To ensure your organization remains on track and maximizes its investment, it's essential to establish clear metrics for success and continuously monitor progress. This chapter explores the critical role of measurement in cloud transformation, guiding you through defining key performance indicators (KPIs), establishing robust feedback loops, and drawing insights from real-world success stories.

Defining Key Performance Indicators (KPIs): Metrics for Success at Every Stage

KPIs are quantifiable measurements that demonstrate the effectiveness of your cloud transformation initiative in achieving its strategic objectives. They offer a clear picture of progress, highlight areas for improvement, and facilitate effective communication of value to stakeholders.

Choosing the Right KPIs

Selecting appropriate KPIs is paramount. Effective KPIs are:

- o Specific: Clearly defined and focused on particular aspects of your transformation.
- o Measurable: Quantifiable and easily tracked.
- o Achievable: Realistic and attainable within your defined timeframe.
- o Relevant: Aligned with your overarching business goals and objectives.

- Time-bound: Associated with specific deadlines or reporting periods.
- Stakeholder Alignment: Metrics for Everyone

Defining KPIs is a collaborative effort. Different stakeholders - executives, IT teams, and end-users - often hold varying priorities. Aligning these perspectives ensures that the metrics you track reflect organizational success comprehensively.

The table below illustrates how diverse priorities can be harmonized into effective, shared KPIs:

Stakeholder	Potential Priority	Potential Conflict	Aligned KPI
IT Team	System Uptime	Cost Optimization	Zero major incidents during new product launches
Marketing	Faster Campaign Rollouts	System Stability	Time to deploy new campaigns without impacting system uptime
Executives	Cost Reduction	Agility	Reduction in TCO while increasing deployment frequency

For instance, while an IT team might prioritize system uptime, marketing could focus on faster campaign rollouts. Bringing these groups together can yield KPIs that balance both perspectives, such as "zero major incidents during new product launches."

KPIs Across the Transformation Lifecycle

KPIs should be dynamically tailored to each stage of your cloud transformation journey:

Planning & Assessment:

- Application Suitability: Percentage of applications assessed and deemed suitable for cloud migration.

- o Cost of Current Infrastructure: Total cost of ownership (TCO) of existing on-premises infrastructure.
- o Projected Cloud Costs: Estimated TCO of the target cloud environment.

Migration:

- o Migration Velocity: Number of applications or workloads migrated per unit of time.
- o Downtime: Total downtime experienced during migration for each application.
- o Migration Cost: Actual cost of migrating each application or workload.

Post-Migration:

- o Cost Optimization: Percentage reduction in infrastructure costs compared to on-premises.
- o Performance Improvement: Percentage increase in application performance (e.g., response time, throughput).
- o Scalability: Ability to dynamically scale resources up or down to meet demand.
- o Availability: Percentage of uptime for cloud-based applications.
- o Security: Number of security incidents or vulnerabilities detected and resolved.
- o User Satisfaction: Feedback from end-users on the performance and usability of migrated applications.

Examples of KPIs:

Common examples of impactful KPIs include:

- o Cost Reduction: A critical area with significant expectations. Key metrics include:
- o Rightsizing Instances: Percentage of instances appropriately sized for their workload, avoiding over-provisioning.

- Spot Instance Utilization: Percentage of workloads running on cost-effective spot instances (AWS) or preemptible VMs (Google Cloud).
- Storage Optimization: Percentage of data stored in cost-effective storage tiers (e.g., cold storage, archive storage).
- Serverless Adoption: Cost savings achieved by using serverless functions or services compared to traditional server-based applications.
- Automation Savings: Cost reduction due to automation of tasks like scaling, backups, and deployments.
- Infrastructure Cost Savings: Overall reduction in operational expenses, including hardware usage, maintenance, and energy consumption.
- Performance Improvement: Increased application response time, improved data processing speed, and enhanced system scalability.
- Agility & Innovation: Faster time-to-market for new products, increased frequency of software releases, and better adaptability to business changes.
- Customer Experience: Reduced customer churn, increased satisfaction, and improved Net Promoter Score (NPS).

Expanding on Agility and Innovation KPIs

Cloud transformation empowers organizations to become more agile and innovative. Agility signifies the ability to respond swiftly to change, while innovation involves developing new products, services, and processes. To truly capture these benefits, it's crucial to measure improvements in agility and innovation, even if they are more challenging to quantify than traditional metrics. These often represent the most significant long-term value. Here's a deeper look at key KPIs in this area:

Deployment Frequency: Measures how often new features, updates, or bug fixes are successfully deployed to production. A higher frequency indicates a more agile development process, allowing for faster delivery of value. Cloud environments, with their automation capabilities and infrastructure-as-code, significantly increase deployment frequency.

Lead Time for Changes: Refers to the time it takes for a change (e.g., a new feature, a bug fix) to move from code commit to successful deployment in production. Shorter lead times enable quicker responses to market demands and user feedback. Cloud practices like continuous integration and continuous delivery (CI/CD) are essential for reducing lead times.

Failure Rate (or Change Failure Rate): Tracks the percentage of deployments that result in a failure, requiring a rollback, hotfix, or other remediation. While failures are inevitable, a lower failure rate indicates a more stable and reliable deployment process. Cloud environments, with their scalability and monitoring tools, help reduce failure rates and minimize impact.

Time to Recover (or Mean Time to Recovery - MTTR): In the event of a failure, this KPI measures how quickly the team can restore service to users. Faster recovery times minimize downtime and disruption, preserving user satisfaction and business continuity. Cloud features like automated backups, disaster recovery services, and infrastructure redundancy contribute to quicker recovery.

Experimentation Rate: Reflects the number of experiments, A/B tests, or proofs of concept the organization can conduct within a given timeframe. A higher experimentation rate fosters a culture of innovation and enables data-driven decision-making. Cloud platforms simplify experimentation with scalable environments for testing and data analytics for measuring results.

The Link Between Cloud Capabilities and Agility/Innovation KPIs

Cloud computing plays a vital role in enabling and enhancing these agility and innovation KPIs:

- Containers and Orchestration (e.g., Docker, Kubernetes): These technologies allow for portable and scalable application packaging and deployment, increasing deployment frequency and reducing lead times.
- Serverless Computing (e.g., AWS Lambda, Azure Functions): Serverless architectures free developers from managing underlying infrastructure, accelerating development cycles and increasing experimentation rates.
- Managed Databases: Cloud-managed databases offer scalability, high availability, and automated maintenance, reducing the operational burden on development teams and allowing for quicker iteration.
- Automation Tools: Cloud platforms provide a wide range of automation tools for CI/CD, infrastructure provisioning, and testing, streamlining development processes, reducing failure rates, and improving time to recover.
- Data Analytics and Machine Learning: Cloud-based data analytics and machine learning services empower organizations to analyze data from experiments, gain insights, and make data-driven decisions, fostering innovation.

Tools for KPI Tracking

Cloud providers offer various integrated tools and services for monitoring and measuring KPIs:

- AWS CloudWatch: Provides real-time insights into resource utilization, application performance, and operational health.
- Azure Monitor: Offers a comprehensive suite of monitoring tools for collecting, analyzing, and acting on telemetry data from your cloud and on-premises environments.
- Google Cloud Monitoring: Delivers customizable dashboards and alerts for monitoring the performance, availability, and uptime of your applications and infrastructure.

Considering Both Direct and Indirect Costs

When evaluating the true impact of cloud optimization, it's essential to look beyond just infrastructure bills. Direct costs, such as compute, storage, and network charges, are the most visible. However, indirect costs—which include operational efficiency, reduced system maintenance, and increased developer productivity—can have an even greater long-term impact on the bottom line.

For example, automating deployments may not drastically cut your infrastructure bill, but it can significantly reduce downtime and free up engineering time. Similarly, adopting serverless or managed services might appear cost-neutral at first glance but can drastically boost team agility and accelerate time to value. By accounting for both types of costs, organizations can make smarter decisions that optimize not just budgets, but overall business outcomes.

The Power of Feedback Loops: Using Data to Refine and Improve Transformation Efforts

Establishing robust feedback loops is essential for continuously improving your cloud transformation initiative. By systematically collecting and analyzing data on your KPIs, you can identify areas where you're exceeding expectations and pinpoint those that require adjustments.

The Feedback Loop Process:

- Gather Data: Collect data on your defined KPIs using monitoring tools, surveys, and other relevant sources.
- Analyze Data: Examine the collected data to identify trends, patterns, and areas for improvement.
- Take Action: Based on your analysis, implement necessary changes to your processes, tools, or strategies.
- Monitor Results: Track the impact of your changes on your KPIs and overall transformation goals.

Benefits of Feedback Loops:

Feedback loops offer numerous advantages, fostering a cycle of continuous improvement:

- Continuous Optimization: Enables ongoing refinement of your cloud environment and processes.
- Increased Efficiency: Helps identify and eliminate bottlenecks and inefficiencies.
- Cost Management: Facilitates better resource allocation and cost control.
- Enhanced Performance: Drives continuous improvement in application performance and user experience.
- Risk Mitigation: Allows for proactive identification and mitigation of potential risks.
- More Detailed Examples of Feedback Loop Implementation

Feedback loops are most effective when grounded in actionable data and tied to meaningful outcomes, spanning across all operational dimensions - cost control, performance, security, and user experience.

Cost Optimization Loop: A finance or DevOps team notices monthly cloud spend exceeding forecasts. Using cost analysis tools (e.g., AWS Cost Explorer, Azure Cost Management), they identify underutilized instances running 24/7. After analysis, the team implements a schedule to automatically shut down non-essential environments during off-hours using automation scripts, resulting in an 18% reduction in monthly cost.

Loop cycle: Cost alert → Usage analysis → Rightsizing or auto-scaling policy update → Monitored savings.

Security Feedback Loop: A security team receives an alert about an open storage bucket exposing sensitive data. Using a tool like Azure Security Center or Google Cloud Security Command Center, they trace the issue to a misconfigured permission set in a recent deployment. Automated policies are then implemented to enforce encryption and private access by default on all future buckets.

Loop cycle: Security alert → Root cause identification → Policy automation → Compliance reporting.

User Experience Feedback Loop: The product team receives a spike in negative feedback via in-app surveys related to slow login times. Using Application Performance Monitoring (APM) tools like Datadog or New Relic, they isolate the issue to a new authentication microservice with inefficient database queries. The service is optimized, tested in staging, and redeployed with improved response times.

Loop cycle: User feedback → APM data → Code optimization → Deployment → Monitoring → User satisfaction follow-up.

How Cloud Tools Enable Effective Feedback Loops

Effective feedback loops rely on robust tooling that spans multiple functional areas:

- Monitoring Tools: Tools like AWS CloudWatch, Azure Monitor, or Google Cloud Operations Suite gather real-time telemetry from infrastructure and applications. They track trends, usage spikes, error rates, and resource bottlenecks—all foundational for initiating feedback cycles.
- Alerting Systems: Systems like PagerDuty, Opsgenie, or native cloud alerts notify teams when KPIs fall outside acceptable thresholds. These alerts should be tied to response playbooks that guide next steps and escalation paths.
- Dashboards: Centralized dashboards (e.g., Power BI, Grafana, Looker) provide visibility into KPIs and trends over time. These help stakeholders visually track success metrics, spot anomalies, and reinforce a data-driven culture.
- Automation Tools: CI/CD pipelines (e.g., GitHub Actions, Azure DevOps, Jenkins) and infrastructure-as-code tools (e.g., Terraform, Pulumi) enable fast remediation. Teams can implement fixes and changes automatically, based on signals

from feedback loops—reducing human intervention and accelerating response times.

Institutionalizing Feedback Loops: The Review Rhythm

Technology alone doesn't ensure continuous improvement; structured review processes are essential to embed feedback loops into organizational culture.

Who Should Be Involved:

- Product Owners and Engineering Leads: Translate data into actionable backlog items.
- IT Operations and DevOps: Oversee system health, cost trends, and platform resilience.
- Security Teams: Review posture, alerts, and incident response.
- Business Stakeholders: Ensure KPIs align with broader organizational outcomes (e.g., customer retention, revenue growth).

How Often to Meet:

- Weekly Tactical Reviews: Focused on recent incidents, optimization opportunities, and sprint retrospectives.
- Monthly KPI Reviews: Evaluate broader trends, assess tooling effectiveness, and plan for systemic changes.
- Quarterly Strategic Reviews: Align metrics with business outcomes, reassess KPI relevance, and fine-tune cloud strategies.

What the Agenda Should Include:

- Review of top KPIs and thresholds crossed.
- Incident summary and follow-up outcomes.
- Cost and resource utilization trends.
- New opportunities for automation or tooling improvement.
- Strategic alignment with business priorities.

Bottom Line: Feedback loops are the operational engine of continuous improvement. When connected across departments, powered by the right tooling, and institutionalized through regular cadence, they transform cloud transformation from a static initiative into a living system - adaptive, accountable, and relentlessly optimized.

Continuous Improvement with AI and Automation

AI and automation are revolutionizing cloud management by making it proactive instead of reactive. Tools powered by AI can identify patterns, predict trends, and automate adjustments to optimize performance and cost. For example, a retailer used AI-driven forecasting to predict traffic surges during holiday sales and scaled their infrastructure in advance, preventing outages and improving customer experience.

Case Studies of Success: Real-world Examples of Businesses that Achieved ROI Through Cloud Transformation

Examining real-world examples of successful cloud transformations provides valuable insights and inspiration for your own journey.

- Netflix: The streaming giant migrated its entire infrastructure to AWS, achieving significant cost savings, improved scalability, and enhanced global reach.
- Airbnb: By leveraging AWS's cloud services, Airbnb scaled its platform to accommodate millions of users worldwide, improving application performance and accelerating innovation.
- Capital One: The financial institution embraced a cloud-first strategy, migrating core banking systems to AWS, enhancing security, and improving customer experience.
- CarMax: The used car retailer migrated its website and core applications to Azure, improving site performance, scalability, and customer experience.

- ASOS: This global online fashion retailer uses Azure to power its e-commerce platform, handling millions of transactions daily, and successfully reduced page load times by 50%.
- Spotify: The music streaming giant uses Google Cloud's data analytics and machine learning capabilities to personalize user experiences, recommend songs, and optimize music discovery.
- Major League Baseball: MLB uses Google Cloud to analyze vast amounts of data, including player statistics, game footage, and fan engagement metrics.
- 20th Century Fox: The film studio leverages Google Cloud's AI and machine learning tools to enhance the movie-making process, including analyzing scripts, predicting audience preferences, and optimizing visual effects rendering.

These case studies demonstrate the diverse applications of cloud platforms across various industries. Each company leveraged the unique strengths of their chosen cloud provider to achieve specific business objectives, from improving customer experience to driving innovation and optimizing operations.

Virtual-World Scenario: A Company's Journey to Measurable ROI and Sustained Success

Let's delve into an example scenario of a company that effectively leveraged KPIs and feedback loops to achieve measurable ROI and sustained success through cloud transformation.

The Challenge:

Acme Manufacturing, a mid-sized company, faced significant challenges with its aging on-premises infrastructure. Frequent downtime, slow application performance, and escalating IT costs hindered operational efficiency and customer satisfaction. Recognizing the need to modernize, Acme Manufacturing sought to improve its IT infrastructure to drive business growth.

The Solution:

Acme Manufacturing decided to migrate its critical applications and data to the cloud, adopting a phased approach. They began with their customer relationship management (CRM) and enterprise resource planning (ERP) systems.

Defining KPIs:

Before embarking on the migration, Acme Manufacturing defined clear KPIs to measure the success of its cloud transformation initiative:

Cost Reduction:

- 20% reduction in IT infrastructure costs within the first year.
- 10% reduction in IT operational expenses within the first year.

Performance Improvement:

- 30% improvement in application response time for CRM and ERP systems.
- 99.99% uptime for critical applications.

Agility & Innovation:

- Reduced time-to-market for new product launches by 20%.
- Increased frequency of software releases from quarterly to monthly.

Establishing Feedback Loops:

Acme Manufacturing implemented a robust monitoring and feedback system to track progress and identify areas for improvement. They leveraged cloud-native monitoring tools to collect real-time data on application performance, resource utilization, and security posture. Regular reviews and analysis of this data enabled Acme Manufacturing to make informed decisions and continuously optimize its cloud environment.

Achieving Success:

By consistently tracking its KPIs and leveraging feedback loops, Acme Manufacturing achieved significant results:

- Cost Optimization: The company exceeded its cost reduction targets, achieving a 25% reduction in IT infrastructure costs and a 15% reduction in operational expenses within the first year.
- Performance Enhancement: Application response time for CRM and ERP systems improved by 35%, surpassing the initial target. The company also achieved its uptime goal of 99.99% for critical applications.
- Increased Agility: The cloud's flexibility enabled Acme Manufacturing to accelerate its product development cycles, reducing time-to-market for new products by 25%. The company also successfully increased its software release frequency from quarterly to monthly.
- Sidebar: When Measurement Is Missing – The APM Wake-Up Call

At Allied Precision Manufacturing (APM), the cloud migration was technically successful - on time, on budget, and fully functional at Go Live. But without clear post-migration KPIs or feedback loops, problems went undetected. Infrastructure costs quietly ballooned. Performance complaints trickled in from Sales and Operations. Months later, what looked like a win required expensive remediation.

Lesson learned: what isn't measured won't improve - and what goes unmeasured can quietly erode the business case.

Sustained Success:

Acme Manufacturing's commitment to measurement and continuous improvement extended beyond the initial migration phase. The company continued to monitor its KPIs, adapt its cloud strategy, and optimize its environment to ensure ongoing success. This proactive approach enabled Acme Manufacturing to maintain its cost savings, performance gains, and agility benefits over the long term.

Key Takeaways:

Acme Manufacturing's success story highlights the importance of:

- Clearly defining KPIs: Setting specific, measurable, achievable, relevant, and time-bound goals.
- Establishing feedback loops: Continuously monitoring, analyzing, and acting on data to drive improvement.
- Embracing a culture of measurement: Making data-driven decisions and prioritizing continuous optimization.

By following these principles, you can significantly increase the likelihood of achieving your cloud transformation objectives and realizing the full potential of the cloud.

Conclusion

Measuring success is not merely a post-migration activity; it's an ongoing process that should be embedded throughout your cloud transformation journey. By defining clear KPIs, establishing effective feedback loops, and learning from real-world examples, you can ensure that your cloud investment delivers measurable ROI and sustained business value. Remember, the cloud is not a magic bullet; it's a powerful tool that requires careful planning, execution, and continuous monitoring to unlock its full potential.

Chapter 12: The Future of Legacy System Transformation

The pace of technological change has accelerated rapidly. For organizations reliant on legacy systems, keeping up isn't just a challenge - it's a mandate for survival. As new technologies emerge, the transformation of legacy systems evolves from a one-time overhaul to an ongoing journey of innovation and agility.

In this chapter, we'll explore the key drivers shaping the future of legacy system transformation, including the role of emerging technologies like AI and machine learning, the shifting responsibilities of the C-Suite, and strategies for staying agile in an unpredictable business environment.

Emerging Technologies: AI, Machine Learning, and Beyond

Emerging technologies are the new architects of legacy system transformation, turning outdated infrastructures into modern, intelligent ecosystems. Here's how they're shaping the future:

Artificial Intelligence (AI): The Brain of Transformation

AI has moved from hype to a practical tool for legacy system enhancement. By integrating AI into legacy systems, businesses can automate repetitive tasks, identify inefficiencies, and predict future needs. AI models can analyze historical data within legacy systems to uncover trends and drive smarter business decisions.

Examples:

- Predictive Maintenance: AI and ML analyze sensor data from legacy equipment to predict failures and optimize maintenance schedules, reducing downtime, improving safety, and extending the lifespan of existing assets.

- Process Automation: AI automates tasks within legacy systems, such as data entry, report generation, or customer onboarding, freeing up human resources.
- Decision Support: AI analyzes data from legacy systems to provide insights that support better decision-making, including identifying trends, forecasting demand, or assessing risks.

A financial services firm used AI to analyze customer transaction histories stored in legacy databases, enabling them to predict customer churn and launch targeted retention campaigns.

Machine Learning (ML): Learning as You Go

ML takes AI a step further by enabling systems to adapt and improve without human intervention. Legacy systems integrated with ML can learn from their data and evolve over time, extending their lifecycle and usefulness.

Use Case:

Manufacturing firms are using ML to predict equipment failures based on historical performance data from legacy systems, reducing downtime and maintenance costs.

The Cloud as a Catalyst

While AI and ML add intelligence, cloud platforms serve as the backbone for transformation. The cloud provides the scalability and processing power required to implement these technologies at scale, while hybrid and multi-cloud strategies offer flexibility for companies still reliant on legacy architectures.

Emerging Cloud Trends:

Serverless Architectures: Minimize infrastructure concerns and focus on functionality. Serverless computing allows developers to execute code without provisioning or managing servers. The cloud provider automatically handles infrastructure scaling, availability, and management. This approach can be particularly beneficial for modernizing specific legacy processes.

Example: Consider a company with a legacy e-commerce platform. The order processing component, built on outdated technology, is slow and difficult to scale during peak demand. By using serverless functions, the company can modernize this process.

When a customer places an order, the platform can trigger a serverless function.

This function can handle tasks such as:

- Validating order details.
- Updating inventory in the legacy database.
- Initiating payment processing through a modern API.
- Sending confirmation emails.

Serverless architecture offers several advantages in this scenario:

- Scalability: The order processing function automatically scales to handle fluctuations in order volume, ensuring smooth operation during peak sales periods.
- Cost-efficiency: The company only pays for the compute time used to execute the function, reducing costs during low-traffic periods.
- Faster Development: Developers can focus on writing code for the order processing logic without managing the underlying infrastructure, accelerating development and deployment.
- Edge Computing: Brings processing closer to the data source, enhancing speed and reducing latency. Instead of sending all data to a centralized cloud for processing, edge computing distributes computing power to devices or local servers at the "edge" of the network, near where the data is generated. This is crucial for applications requiring real-time responses.

Example: Consider a manufacturing company with a legacy factory automation system. The system relies on sensors that collect data on equipment performance, but the data is sent to a central server for analysis, causing delays in identifying and responding to critical issues. By implementing edge computing, the company can modernize this process:

- Edge devices or local servers are deployed on the factory floor.

These edge devices can:

- Collect and process sensor data in real-time.
- Identify anomalies or deviations from normal operating parameters.
- Trigger immediate alerts or automated responses to prevent equipment failures.
- Send only essential summary data to the central server for long-term analysis and reporting.

Edge computing offers several advantages in this scenario:

- Reduced Latency: Processing data locally minimizes the time it takes to respond to critical events, preventing damage or downtime.
- Increased Reliability: Edge computing can continue to operate even if connectivity to the central cloud is disrupted.
- Bandwidth Optimization: Only sending essential data to the cloud reduces bandwidth consumption and network congestion.

The Evolving Role of the C-Suite: From Decision-Makers to Transformation Leaders

Successfully navigating this technological shift requires a corresponding evolution in leadership. The C-Suite must adapt its role to effectively guide and champion legacy system transformation.

The C-Suite has traditionally been responsible for strategic decisions, but in today's rapidly evolving tech landscape, their role is expanding into that of transformation leaders. Here's how their responsibilities are shifting:

Championing Change

Leaders like the CIO and CTO are no longer gatekeepers of IT - they're catalysts for business transformation. The CEO, too, is increasingly required to understand how technology aligns with overarching business objectives.

Key Focus Areas:

- Tech Literacy: C-Suite leaders must understand enough about emerging technologies to drive informed decisions.
- Cross-Functional Collaboration: Successful transformations depend on breaking down silos between IT and other business units.
- Risk Management: Balancing innovation with cybersecurity and compliance concerns.

Encouraging a Culture of Innovation

Transformation starts at the top, and the C-Suite plays a critical role in fostering a culture that embraces change. Leaders must set the tone for risk-taking and adaptability while ensuring employees are equipped to thrive in a tech-enabled workplace.

Example: A global retail company created an "Innovation Board" led by C-level executives, which empowered cross-departmental teams to propose and pilot new technologies for legacy system improvements.

Navigating Talent and Skills Gaps

The C-Suite must also address the growing talent gap in emerging tech. Legacy systems often rely on outdated programming languages and workflows, making it critical to upskill existing staff or bring in fresh talent.

Strategies:

- Partner with academic institutions to create specialized training programs.
- Invest in tools that enable citizen developers to participate in innovation.
- Low-Code/No-Code Solutions: Empower existing employees to contribute to innovation and development, even without deep coding skills.
- Reskilling Initiatives: Implement reskilling programs to train the workforce on new technologies relevant to legacy system modernization.
- Staying Agile: Innovating in an Ever-Changing Business Landscape
- Agility is no longer just a buzzword, it's a survival strategy. For businesses reliant on legacy systems, agility means finding ways to innovate without losing stability.
- Phased Transformation: Bite-Sized Progress

Rather than attempting a complete overhaul, businesses can embrace a phased transformation approach, addressing specific system components over time. This approach reduces risk while delivering incremental value.

Case Study: A logistics company modernized its warehouse management system in phases, starting with data migration to the cloud. Each phase delivered measurable ROI, enabling reinvestment in subsequent upgrades.

While a phased approach offers clear advantages, some organizations still attempt full-scale overhauls with mixed results.

Cautionary Tale: One virtual example is Assurance Health, which opted to replace its entire claims processing system in a single 12-

month push. Lacking a modular strategy or iterative testing, the go-live introduced system-wide outages, record mismatches, and prolonged staff retraining. The initiative ultimately took double the time and triple the budget to stabilize. This underscores why bite-sized transformation isn't just a low-risk tactic - it's often the only path to sustainable modernization.

Adopting Flexible Architectures

Microservices and APIs have become essential tools for enhancing agility. These technologies allow legacy systems to integrate new solutions, enabling organizations to innovate without abandoning their foundational systems.

Example: A healthcare provider implemented an API-based interface to connect its legacy patient management system with a modern telemedicine platform, enabling a seamless patient experience.

Continuous Improvement: The Feedback Loop of Agility

Organizations must embed continuous improvement into their DNA by establishing feedback loops that analyze performance and inform future actions. Regular assessments of technology, processes, and outcomes ensure that innovation remains aligned with business goals.

Tip: Use OKRs (Objectives and Key Results) to measure progress against transformation goals, ensuring agility doesn't compromise long-term objectives.

The Composable Enterprise: A New Paradigm for Legacy Modernization

The concept of the "composable enterprise" is gaining traction as a way for organizations to build more modular and adaptable systems. This approach involves breaking down business processes into smaller, self-contained components called packaged business capabilities (PBCs). These PBCs can be combined and recombined as needed to meet changing business requirements.

For legacy modernization, the composable enterprise offers a compelling path forward. Instead of ripping and replacing entire systems, companies can incrementally replace or augment legacy components with modern, cloud-based PBCs. This allows for a more agile and iterative approach to transformation, reducing risk and delivering faster time-to-value.

Ethical Considerations in AI-Driven Legacy Transformation

As AI plays an increasing role in legacy system transformation, it's essential to address the ethical considerations that arise.

Bias in AI Models: Legacy systems often contain biased data that reflects historical inequalities or prejudices. When using AI/ML to analyze this data, it's crucial to identify and mitigate these biases to avoid perpetuating or amplifying discriminatory outcomes.

Explainability and Transparency: In sensitive areas like healthcare or finance, decisions made by AI need to be understood and trusted. It's important to use transparent and explainable AI models so that humans can understand how decisions are being made and intervene if necessary.

Conclusion

The future of legacy system transformation is dynamic, driven by emerging technologies, the expanding role of leadership, and the necessity for agility. By leveraging AI, machine learning, and the cloud, businesses can breathe new life into their systems while staying competitive.

The C-Suite must step into the role of transformation leaders, fostering a culture of innovation and bridging skills gaps. Organizations must embrace agility as a core principle, ensuring they remain adaptable in an ever-changing landscape. Transformation is no longer a one-time project; it's an ongoing journey. By embracing these strategies, businesses can ensure their legacy systems don't just survive, they thrive in the future of work.

To navigate this evolving landscape successfully, organizations must:

- Embrace continuous learning and experimentation to stay ahead of technological advancements.
- Cultivate a proactive and adaptable mindset when approaching legacy system transformation.
- Prioritize ethical considerations when leveraging AI and other emerging technologies.

By taking these steps, businesses can ensure their legacy systems remain valuable assets in the digital age.

Chapter 13: A Call to Action

As the final chapter of this journey, let's pause and reflect on where we've been. Cloud transformation isn't just about technology - it's about unlocking possibilities, driving innovation, and building a resilient future. By now, you've explored the strategies, tools, and frameworks necessary to modernize legacy systems and seize the opportunities of a rapidly evolving digital landscape.

This chapter is your call to action. It's time to take the insights and inspiration you've gathered and translate them into meaningful steps. Whether you're just starting your transformation journey or refining your existing approach, the future is yours to shape.

Taking the First Step

Every transformation begins with a single, deliberate step. That step might look different depending on where your organization stands today, but the key is to act decisively. Here's how to set the wheels in motion:

Assess Your Starting Point

Before charting your course to the cloud, it's crucial to understand your current landscape. This involves a comprehensive assessment of your existing IT infrastructure, applications, and data. Think of it as taking inventory – you need to know what you have before you can decide what to keep, what to discard, and what to transform.

Key Areas to Analyze:

- Infrastructure: Evaluate your servers, network devices, and storage systems. Identify their age, capacity, performance limitations, and maintenance costs.
- Applications: Catalog your business applications, their dependencies, and their current performance levels. Assess their suitability for cloud migration and identify potential challenges.
- Data: Analyze your data storage, its volume, types, and sensitivity. Evaluate data governance policies and compliance requirements.
- Security: Review your current security posture, including access controls, threat detection mechanisms, and data protection measures.
- Skills: Assess your team's skills and expertise in cloud technologies. Identify any knowledge gaps that need to be addressed through training or recruitment.

Cloud Readiness Assessment:

To streamline this process, consider conducting a formal cloud readiness assessment. Several tools and services are available to help you evaluate your organization's preparedness for cloud adoption. These assessments can provide valuable insights into your strengths, weaknesses, and areas for improvement.

Actionable Step: Assemble a cross-functional team with representatives from IT, business units, and security to conduct this assessment. This team should be responsible for mapping out your IT landscape, including legacy systems, integration points, and potential bottlenecks.

Define Clear Goals

Success is impossible without a clear vision. Establish objectives that align with your business strategy, and ensure every stakeholder understands the destination.

- Actionable Step: Use the SMART framework (Specific, Measurable, Achievable, Relevant, Time-bound) to define your cloud transformation goals.

Build Momentum Through Quick Wins

A successful transformation journey relies on building confidence. Focus on initiatives that deliver tangible results early, such as migrating non-critical workloads to the cloud or automating repetitive processes.

- Actionable Step: Pilot a small-scale project, like migrating a single department's application, to demonstrate value and refine your approach.

Commit to Continuous Learning

Cloud transformation is not a one-and-done process. Commit to learning from every phase, iterating your strategy as new challenges and opportunities arise.

- Actionable Step: Create a roadmap for periodic reviews, measuring progress against your KPIs and making necessary adjustments.

Leading with Vision

While these initial steps lay the groundwork for transformation, strong leadership is essential to guide the organization through the journey and ensure its long-term success. The most successful transformations are driven by leaders who view change as an opportunity rather than a challenge. Visionary leadership inspires teams to embrace uncertainty and move forward with confidence.

Empowering Teams to Succeed

Empower your teams by fostering a culture of innovation and providing the tools they need to succeed. Encourage collaboration, reward creativity, and invest in skill development to bridge gaps in expertise.

- Example: A manufacturing company empowered its employees to become "cloud champions" by offering training programs and recognizing innovative ideas for leveraging new tools.

Inspiring Ownership at Every Level

Transformation isn't just an IT initiative, it's an organization-wide effort. Empower individuals at all levels to take ownership of their role in the journey.

- Actionable Step: Establish transformation ambassadors within each department to serve as advocates and facilitators for change.

Communicating the "Why" Behind the Journey

A clear and compelling narrative is essential to rallying support. Communicate the purpose of your transformation—not just what you're doing but why it matters.

- Tip: Use stories to illustrate the impact of change, whether it's improving customer satisfaction, enhancing operational efficiency, or enabling innovation.

Cloud Transformation: A Catalyst for Reinvention

The cloud isn't just a tool; it's a catalyst for reinvention. It has the power to redefine how businesses operate, compete, and grow. By embracing cloud transformation, organizations position themselves to thrive in an increasingly connected, data-driven world.

The Role of Cloud Transformation

Cloud transformation isn't simply about migrating applications or reducing costs, it's about shaping the future. It enables businesses to:

- Adapt: Stay agile in a rapidly changing environment.
- Innovate: Leverage emerging technologies to create new value.
- Thrive: Deliver exceptional experiences to customers, employees, and stakeholders.

Your Role in the Journey

As a leader or decision-maker, your role is pivotal. You have the power to inspire action, create alignment, and turn potential into reality. The steps you take today will define not just the trajectory of your organization but also its ability to compete and excel in the future.

Final Encouragement: Don't wait for the perfect moment. The best transformations start with imperfect beginnings and grow stronger with every iteration. Take the first step, lead with vision, and embrace the opportunities ahead.

Conclusion: The Future is Yours

Ultimately, cloud transformation is about empowering people – employees, customers, and stakeholders – to achieve more. By embracing this people-centric approach, organizations can unlock the full potential of the cloud and create a future where technology serves as a catalyst for human progress.

This book has been your guide, but the journey ahead is uniquely yours. Cloud transformation is not the destination, it's the road that leads to growth, innovation, and resilience. With the tools, strategies, and inspiration from this journey, you're equipped to shape the future of your organization and leave a lasting impact.

Go forth and transform. The future is waiting.

Section A1 - Top Criteria for Evaluating a Public Cloud

Whether you're a seasoned IT pro or just dipping your toes into the cloud, these are the essentials you can't afford to overlook.

Compute Power

Think of this as the horsepower under the hood of your cloud machine. A solid provider will offer a buffet of virtual machines (VMs) with different processing power, memory, and storage to suit everything from lightweight apps to heavy-duty data crunching. You need versatility here—it's like having the right tool for every job.

Storage Options

Data's the lifeblood of your business, so where you stash it matters. Look for a mix of options: object storage for big, unstructured data dumps, block storage for high-performance needs, and file storage for traditional setups. The key is having a solution that fits whether you're archiving old data or zipping through real-time transactions.

Networking Capabilities

Imagine your cloud setup as a bustling city—you need highways, intersections, and some good ol' traffic cops. Virtual networks, secure on-ramps to your on-premises systems, and features like load balancing and firewalls are your infrastructure. A cloud provider with solid networking chops keeps the traffic flowing smoothly and securely.

Database Services

Whether you're running a traditional relational database, dabbling in NoSQL, or need a full-blown data warehouse, you want someone who handles the heavy lifting. Managed database services let you focus on using your data, not babysitting it. This is like having a reliable sous chef in your kitchen—you set the recipe, and they handle the grunt work.

Security Features

This one's non-negotiable, my friends. Your provider needs top-notch security—think encryption, access controls, and threat detection. It's like locking the front door, installing an alarm, and having a guard dog all rolled into one. Your data deserves no less.

Flexible Pricing Models

Budgets matter. A good provider lets you pay as you go, lock in savings with reserved instances, or snag deals with spot instances. It's about matching costs to your actual needs, not overpaying for stuff you don't use. Think of this as picking the right data plan for your cell phone—use it wisely, and you'll save a bundle.

Global Reach

If your business spans the globe, your cloud provider should too. Data centers in multiple regions mean you can deploy applications closer to your users, cutting down on lag time and keeping everyone happy. High availability and reduced latency? That's what I call a win-win.

Every cloud provider has its strengths and weaknesses. Your job is to stack these criteria against your business's unique needs. Kick the tires, do the homework, and choose the one that's going to help you not just survive, but thrive.

Section A2 – Example Disaster Recovery Plan for a Cloud Environment

Let's imagine a company called "Acme Corp" that sells widgets online. All their critical applications, like their website, order processing system, and customer database, are hosted on a public cloud platform like AWS (Amazon Web Services). Here's how their disaster recovery plan might look:

1. Data Backup and Replication:

Regular Backups: Acme Corp backs up their entire system, including databases, applications, and configurations, at regular intervals (e.g., daily or hourly). These backups are stored in a different geographic location than their primary cloud region.

Data Replication: They use real-time data replication to create a mirror copy of their critical data in a secondary cloud region. This ensures that even if their primary region goes down, they have an up-to-date copy of their data ready to go.

2. Infrastructure Redundancy:

Multiple Availability Zones: Acme Corp deploys their applications across multiple availability zones within their primary region. Availability zones are isolated locations within a region that are designed to be independent of each other. If one zone goes down, the others can continue operating.

Pilot Light Environment: They maintain a "pilot light" environment in their secondary region. This is a minimal version of their production environment that includes core components and critical data. In the event of a disaster, they can quickly "ignite" this environment and scale it up to handle production traffic.

3. Failover Mechanisms:

Automated Failover: Acme Corp has implemented automated failover mechanisms that detect outages in their primary region and automatically redirect traffic to their secondary region. This ensures minimal downtime and disruption to their business.

DNS Routing: They use DNS (Domain Name System) routing to direct traffic to the appropriate region. In case of a disaster, they can update their DNS records to point to their secondary region.

4. Testing and Recovery Procedures:

Regular Disaster Drills: Acme Corp conducts regular disaster recovery drills to test their plan and ensure that their systems and teams are prepared for an actual disaster.

Documented Recovery Procedures: They have detailed documentation outlining the steps to take in the event of a disaster, including contact information for key personnel and escalation procedures.

5. Monitoring and Alerting:

Real-time Monitoring: Acme Corp uses monitoring tools to track the health and performance of their cloud environment. They receive alerts in case of any issues or anomalies.

Incident Response Team: They have a dedicated incident response team that is trained to handle disaster scenarios and coordinate recovery efforts.

In the event of a disaster, such as a major outage in their primary cloud region, Acme Corp can quickly recover their operations by:

Failing over to their secondary region: Their automated failover mechanisms will redirect traffic to their pilot light environment in the secondary region.

Scaling up their infrastructure: They will scale up their resources in the secondary region to handle the increased traffic.

Restoring data from backups: If necessary, they will restore data from their backups to ensure data consistency.

By implementing these disaster recovery measures, Acme Corp can minimize downtime, protect their data, and ensure business continuity in the face of unexpected events.

Section A3 – Example Cloud Adoption Framework (CAF)

Legacy System Modernization Framework (LSMF)

A Strategic Guide for Cloud Transformation

Version: 1.0
Author: Joel Steven
Date: June 2025

1. Introduction

Legacy systems are the backbone of many enterprises, yet their aging infrastructure and technical debt often hinder growth, innovation, and agility. This Legacy System Modernization Framework (LSMF) provides a structured, phased approach to modernizing legacy environments using cloud solutions. It enables organizations to assess, plan, adopt, and optimize their cloud journey while mitigating risks associated with large-scale transformation.

This framework is designed to help C-suite executives, IT leaders, and business stakeholders balance operational continuity with modernization initiatives.

2. Framework Principles

The LSMF is guided by four key principles:

Business Alignment – Cloud adoption must align with strategic goals, ensuring measurable ROI.

Incremental Modernization – Phased transition strategies (e.g., Rehost, Refactor, Rearchitect, Rebuild) reduce risk.

Security & Compliance First – Data protection, governance, and regulatory adherence remain top priorities.

Optimized Cost & Performance – Cost-effective cloud solutions must enhance system efficiency and scalability.

3. Phases of Modernization

3.1 Assess

Objective: Evaluate the current state of legacy systems, dependencies, and business impact.

Technical Debt Analysis – Identify outdated architectures, unsupported technologies, and scalability constraints.

Cloud Readiness Assessment – Determine which workloads can be migrated, modernized, or replaced.

Stakeholder Alignment – Gain executive buy-in by mapping modernization to business objectives.

3.2 Plan

Objective: Define a strategic roadmap for modernization.

Cloud Strategy Selection – Choose between Hybrid Cloud, Multi-Cloud, or Full Cloud Adoption based on business needs.

Workload Prioritization – Rank applications based on complexity, risk, and value.

Security & Compliance Planning – Implement Zero Trust Architecture, data governance, and industry compliance (e.g., HIPAA, GDPR).

Change Management & Training – Prepare teams for process transformation and cloud-native skill development.

3.3 Adopt

Objective: Execute cloud migration and system transformation.

Migration Pathways:

- Rehost (Lift & Shift) – Migrate workloads with minimal changes.
- Refactor (Optimize for Cloud) – Modify applications for improved scalability and cost-efficiency.
- Rearchitect (Cloud-Native Transformation) – Rebuild using microservices and serverless architectures.
- Replace (SaaS Adoption) – Transition legacy applications to modern SaaS alternatives.
- Pilot Testing & Validation – Implement proof-of-concept migrations before full deployment.

3.4 Optimize & Govern

Objective: Continuously improve system efficiency, cost-effectiveness, and compliance post-migration.

Performance & Cost Optimization – Use FinOps principles to optimize cloud spend.

Security Posture Monitoring – Deploy SIEM tools for real-time threat detection.

Operational Excellence – Implement DevOps & SRE practices for continuous deployment and reliability.

Business Value Measurement – Track KPIs such as cost reduction, system uptime, and user adoption rates.

4. Tools & Methodologies

4.1 Cloud Adoption Models

Hybrid Cloud Strategy – Balancing on-prem and cloud infrastructure.

Multi-Cloud Approach – Leveraging multiple cloud providers for redundancy and optimization.

Serverless & Edge Computing – Enhancing agility with next-gen cloud-native architectures.

4.2 Governance & Compliance Frameworks

- NIST Cybersecurity Framework
- CIS Cloud Security Benchmarks
- ISO 27001: Cloud Security Standards

4.3 Technology Stack

Compute: Azure VMs, AWS EC2, Google Cloud Compute Engine

Data Management: Azure Synapse, AWS Redshift, Google BigQuery

AI & Automation: Azure Cognitive Services, AWS AI/ML, Google Vertex AI

5. Case Study: Allied Precision Manufacturing (APM)

To illustrate the LSMF in action, this section presents the modernization journey of APM, a mid-sized industrial manufacturing firm struggling with legacy ERP, siloed data systems, and high maintenance costs.

Challenges:

- Legacy on-prem ERP (ProcessPro) lacked real-time analytics.
- High maintenance costs and security vulnerabilities.
- Manual workflows delayed production and supply chain visibility.

Solution Approach:

- Rehosted ERP & CRM to Azure Virtual Machines for quick cloud enablement.
- Refactored Data Infrastructure using Azure Synapse & Power BI for real-time analytics.
- Implemented AI-Driven Predictive Maintenance to reduce downtime.

Outcomes:

- 35% cost savings from infrastructure consolidation.
- Improved decision-making with real-time manufacturing insights.
- Faster production cycles due to automated workflows.

6. Conclusion & Next Steps

Modernizing legacy systems is not a one-time event but a continuous process. Organizations must strategize, execute, and refine their cloud adoption to stay competitive. By following the Legacy System Modernization Framework (LSMF), enterprises can reduce technical debt, enhance business agility, and future-proof their IT landscape.

For organizations embarking on this journey, the next steps include:
✓ Conducting a Cloud Readiness Workshop
✓ Engaging Stakeholders for Business Alignment
✓ Developing a Phased Modernization Plan

Section A4 – Issues Surrounding Non-optimized Applications in the Cloud

Migrating applications to the cloud via a 'lift-and-shift' approach offers a rapid pathway to cloud adoption. However, without optimizing these applications for cloud-native capabilities, organizations risk encountering a range of significant disadvantages. This section outlines the critical challenges, including financial inefficiencies, operational stagnation, technological debt, and performance limitations, that arise from deploying non-optimized applications in a cloud environment.

Problems Arising from Non-Optimized Lift-and-Shift:

Financial Inefficiency:

The pay-as-you-go model of the cloud becomes a liability rather than an asset. Without optimization, you'll over-provision resources, leading to inflated and uncontrolled operational expenses. This stems from the inability of legacy applications to dynamically scale and utilize resources efficiently.

Operational Stagnation:

The promise of cloud automation and orchestration is lost. Legacy applications require manual management, hindering agility and slowing down deployment cycles. This also limits the use of modern cloud security tools and practices, which can increase security vulnerabilities.

Technological Debt and Innovation Blockage:

Simply relocating applications without modernization transfers existing technical debt to the cloud. This debt acts as a barrier to adopting cloud-native technologies, such as serverless functions and containerization, effectively stifling innovation and limiting the ability to modernize.

Performance and Scalability Mismatch:

Legacy applications are not designed for the distributed architecture of the cloud. This leads to performance bottlenecks, latency issues, and an inability to automatically scale resources in response to fluctuating demand.

www.ingramcontent.com/pod-product-compliance
Lightning Source LLC
Chambersburg PA
CBHW072348200326
41519CB00015B/3696